Risk-Based Monitoring and Fraud Detection in Clinical Trials Using JMP® and SAS®

Richard C. Zink

SAS® Documentation

The correct bibliographic citation for this manual is as follows: Zink, Richard C. 2014. *Risk-Based Monitoring and Fraud Detection in Clinical Trials Using JMP® and SAS®*. Cary, North Carolina: SAS Institute Inc.

Risk-Based Monitoring and Fraud Detection in Clinical Trials Using JMP® and SAS®

Copyright © 2014, SAS Institute Inc., Cary, NC, USA

ISBN 978-1-61290-991-2

All rights reserved. Produced in the United States of America.

For a hardcopy book: No part of this publication may be reproduced, stored in a retrieval system, or transmitted, in any form or by any means, electronic, mechanical, photocopying, or otherwise, without the prior written permission of the publisher, SAS Institute Inc.

For a Web download or e-book: Your use of this publication shall be governed by the terms established by the vendor at the time you acquire this publication.

The scanning, uploading, and distribution of this book via the Internet or any other means without the permission of the publisher is illegal and punishable by law. Please purchase only authorized electronic editions and do not participate in or encourage electronic piracy of copyrighted materials. Your support of others' rights is appreciated.

U.S. Government Restricted Rights Notice: The Software and its documentation is commercial computer software developed at private expense and is provided with RESTRICTED RIGHTS to the United States Government. Use, duplication or disclosure of the Software by the United States Government is subject to the license terms of this Agreement pursuant to, as applicable, FAR 12.212, DFAR 227.7202-1(a), DFAR 227.7202-3(a) and DFAR 227.7202-4 and, to the extent required under U.S. federal law, the minimum restricted rights as set out in FAR 52.227-19 (DEC 2007). If FAR 52.227-19 is applicable, this provision serves as notice under clause (c) thereof and no other notice is required to be affixed to the Software or documentation. The Government's rights in Software and documentation shall be only those set forth in this Agreement.

SAS Institute Inc., SAS Campus Drive, Cary, NC 27513-2414.

Printing 1, July 2014

SAS provides a complete selection of books and electronic products to help customers use SAS® software to its fullest potential. For more information about our offerings, visit support.sas.com/bookstore or call 1-800-727-3228.

SAS® and all other SAS Institute Inc. product or service names are registered trademarks or trademarks of SAS Institute Inc. in the USA and other countries. ® indicates USA registration.

Other brand and product names are trademarks of their respective companies.

Contents

About This Book vii
About the Author xi
Acknowledgments xiii

Chapter 1 • Introduction 1
    1.1 Overview 1
    1.2 Topics Addressed in This Book 3
    1.3 The Importance of Data Standards 8
    1.4 JMP Clinical 9
    1.5 Clinical Trial Example: Nicardipine 13
    1.6 Organization of This Book 15
    References 15

Chapter 2 • Risk-Based Monitoring: Basic Concepts 19
    2.1 Introduction 19
    2.2 Risk Indicators 22
    2.3 Geocoding Sites 49
    2.4 Reviewing Risk Indicators 53
    2.5 Final Thoughts 73
    References 73
    Appendix 75

Chapter 3 • Risk-Based Monitoring: Customizing the Review Experience 79
    3.1 Introduction 79
    3.2 Defining Alternate Risk Thresholds and Actions 80
    3.3 Performing Additional Statistical and Graphical Analyses 95
    3.4 Creating JMP Scripts and Add-Ins 108
    3.5 Final Thoughts 124
    References 124

Chapter 4 • Detecting Fraud at the Clinical Site 127

4.1 Introduction 127
4.2 Study Visits 129
4.3 Measurements Collected at the Clinical Site 144
4.4 Multivariate Analyses 164
4.5 Final Thoughts 175
References 176

Chapter 5 • Detecting Patient Fraud 179
5.1 Introduction 179
5.2 Initials and Birthdate Matching 180
5.3 Hierarchical Clustering of Pre-Dosing
    Covariates Across Clinical Sites 185
5.4 Review Builder: Quality and Fraud 192
5.5 Final Thoughts 197
References 198

Chapter 6 • Snapshot Comparisons 199
6.1 Introduction 199
6.2 Domain Keys 200
6.3 Review Flags 212
6.4 Adding and Viewing Notes 217
6.5 Using Review Flags 220
6.6 Final Thoughts 237
References 238

Chapter 7 • Final Thoughts 239
A Work in Progress 239
Stay in Touch 240

Recommended Reading 241
Index 243

About This Book

Purpose

There is increasing literature and discussion on the need of the pharmaceutical industry to improve efficiency and reduce costs through the use of centralized monitoring techniques. So-called risk-based monitoring (RBM) makes use of central computerized review of clinical trial data and site metrics to determine if and when clinical sites should receive more extensive quality review through on-site monitoring visits. There are many benefits to RBM beyond cost: Statistical and graphical checks can determine the presence of outliers or unusual patterns in the data, comparisons can be made between sites to assess performance and identify potentially fraudulent data or miscalibrated or faulty equipment, and issues can be identified and resolved while the trial is ongoing. As stated in the recent FDA guidance on the topic, RBM forces the sponsor to take a more proactive approach to quality through a well-defined protocol and sufficient training and communication, and by highlighting those data most important to patient safety and the integrity of the final study results. There is further emphasis on flexibility and adaptability: responding to issues as they arise to prevent further problems, with the willingness to put additional safeguards in place or refine procedures to address shortcomings as the trial is ongoing.

The primary goal of this book is to describe a practical implementation of JMP and SAS for the centralized monitoring of clinical trials. This includes RBM analyses by defining risk indicators and their corresponding thresholds, methods to identify patient- and investigator-perpetrated fraud, and other visual and analytical techniques to enhance safety and quality reviews. A second but equally important goal is to promote these methodologies and establish a dialogue with practitioners for continual improvement of analytical techniques and software. Controlling clinical trial costs through centralized monitoring is important, as these expenses are ultimately passed on to the consumer; reining in costs is one way to keep drugs affordable for the people who need them. Further, we'll see repeatedly throughout the book that centralized monitoring meets the needs for Good Clinical Practice—protecting the well-being of trial participants, and maintaining a high level of data quality to ensure the validity and integrity of the final analysis results. While this book is written from the point of view of drug development, the methods described are equally applicable to clinical trials of biologics (i.e., vaccines) or medical devices.

Is This Book for You?

Any clinical trial is the result of the efforts of a diverse team. This team includes clinicians, statisticians, data managers, programmers, regulatory associates, and monitors, to name a few. Every team member has a role to play in the review of trial data, and each individual brings a unique skill set important for understanding patient safety, protocol adherence, or data insufficiencies that can affect the final analysis, clinical study report, and subsequent regulatory review and approval. Given the interactive, dynamic, and graphical nature of JMP Clinical, it is my sincere hope that any individual from the clinical trial team can make use of this book and the examples contained within to streamline, accelerate, and enrich their reviews of clinical trial data. Though JMP Clinical makes use of SAS, it is neither expected nor required that the user have any SAS programming experience.

Prerequisites

Users familiar with JMP can expect a similar experience with JMP Clinical; analyses and reports are available from a menu-driven, point-and-click interface. Basic knowledge of JMP is useful (though not required) and can be obtained through **Tutorials** in the **Help** menu, the PDF books **Discovering JMP** and **Using JMP** under **Help > Books** or by referring to *JMP Essentials, Second Edition* by Curt Hinrichs and Chuck Boiler. This book focuses heavily on the operational and data aspects of clinical trials. Familiarity in these areas will make for a more pleasant reading experience. Data standards from the Clinical Data Interchange Standards Consortium (CDISC) are a requirement to use JMP Clinical and are discussed throughout this book. Some familiarity with these standards is useful, though they are addressed at a basic level.

About the Examples

Software Used to Develop the Book's Content

JMP Clinical 5.0, the software version that is the focus of this book, is a combination product that includes JMP 11.1 and SAS 9.4 M1 for Base SAS, SAS/GRAPH, and SAS/ACCESS to PC files, and SAS analytics releases 12.3 for SAS/IML, SAS/STAT, and SAS/Genetics. Also included are macros written in SAS and the JMP Scripting Language (JSL) that generate the various analyses and reports that are specific to JMP Clinical.

The most straightforward answer, however, is simply JMP Clinical 5.0.

Example Code and Data

Data for the clinical trial of Nicardipine is included and shipped with JMP Clinical. Users should have immediate access to this data.

You can access additional supporting files for this book by linking to its author page at http://support.sas.com/publishing/authors. Select the name of the author. Then, look for the cover thumbnail of this book, and select Example Code and Data to display the SAS programs that are included in this book.

For an alphabetical listing of all books for which example code and data is available, see http://support.sas.com/bookcode. Select a title to display the book's example code.

If you are unable to access the code through the Web site, send an e-mail to saspress@sas.com.

Output and Graphics Used in This Book

All output was generated using JMP Clinical 5.0 and captured using FullShot 9.5 Professional.

Additional Resources

SAS offers you a rich variety of resources to help build your SAS skills and explore and apply the full power of SAS software. Whether you are in a professional or academic setting, we have learning products that can help you maximize your investment in SAS.

Keep in Touch

We look forward to hearing from you. We invite questions, comments, and concerns. If you want to contact us about a specific book, please include the book title in your correspondence.

To Contact the Author through SAS Press

By e-mail: saspress@sas.com

Via the Web: http://support.sas.com/author_feedback

SAS Books

For a complete list of books available through SAS, visit http://support.sas.com/bookstore.

Phone: 1-800-727-3228

Fax: 1-919-677-8166

E-mail: sasbook@sas.com

SAS Book Report

Receive up-to-date information about all new SAS publications via e-mail by subscribing to the SAS Book Report monthly eNewsletter. Visit http://support.sas.com/sbr.

Publish with SAS

SAS is recruiting authors! Are you interested in writing a book? Visit http://support.sas.com/saspress for more information.

About the Author

Richard C. Zink is a Principal Research Statistician Developer in the JMP Life Sciences division at SAS Institute. He is currently a developer for JMP Clinical, an innovative software package designed to streamline the review of clinical trial data. He joined SAS in 2011 after eight years in the pharmaceutical industry, where he designed and analyzed clinical trials for patients diagnosed with chronic hepatitis B infection, chronic myeloid leukemia, glaucoma, dry eye disease, blepharitis, or cystic fibrosis; he also participated in U.S. and European drug submissions and in two FDA advisory committee hearings. When not actively engaged in clinical development responsibilities, he supported non-clinical development, pharmaceutical sciences, and sales and marketing activities.

Richard is an active member of the Biopharmaceutical Section of the American Statistical Association, the Drug Information Association, and Statisticians in the Pharmaceutical Industry. He is currently the Statistics Section Editor for *Therapeutic Innovation & Regulatory Science* (formerly *Drug Information Journal*). He is a frequent speaker at workshops and scientific meetings and has lectured for courses in statistics and clinical trials. His research interests include the analysis of pre- and post-market adverse events, subgroup identification for patients with enhanced treatment response, and risk-based monitoring and fraud detection in clinical trials.

Richard holds a Ph.D. in Biostatistics from the University of North Carolina at Chapel Hill and has more than 20 years of SAS programming experience. This is his first book.

Acknowledgments

First conceived in a fit of inspiration (read: madness) while mowing my front lawn, this book would not exist without the contributions and support of many individuals.

Thanks to Stacey Hamilton, Cindy Puryear, and Shelley Sessoms at SAS Press for their excitement and encouragement. Many thanks to the reviewers for their insightful comments that improved the content and clarity of this book, and to Brenna Leath, who made the text consistent throughout.

Thanks to the JMP Life Sciences team for its key role in the development, documentation, testing, and promotion of JMP Clinical: Wenjun Bao, Tzu-Ming Chu, John Cromer, Drew Foglia, Lili Li, Geoffrey Mann, Kelci Miclaus, Thomas Pedersen, Anisa Scott, Susan Shao, Luciano Silva, and Russ Wolfinger.

Thanks to JMP for its support of the Life Sciences, especially John Sall, Laura Archer, Jim Borek, Anne Bullard, Jian Cao, Shannon Conners, Ian Cox, Wu Deng, Xan Gregg, Bob Hickey, Jordan Hiller, Kiyomi Kato, Chris Kirchberg, Laura Lancaster, Sheila Loring, Bernard McKeown, Arati Mejdal, Adam Morris, Valérie Nedbal, Chung-Wei Ng, John Ponte, Doug Robinson, Julie Stone, Nala Sun, Kyoko Takenaka, Walter Teague, Nan Wu, and Da Zhang. Thanks to SAS, especially Wendy Czika, Anders Larsen, Gene Lightfoot, Jamie Powers, Laurie Rose, Wayne Saraphis, and Diana Witt.

Special thanks to Andy Lawton for his generous feedback and suggestions. Thanks to Shrikant Bangdiwala, Jon Blatchford, Nancy Geller, and Gary Kennedy for sharing their insights, and thanks to those individuals too numerous to mention for (repeatedly) asking, "Can't you just show me what's new?"

Thanks to my wife, Sarah, for the love and support she gives me in everything I do. Thanks also for the edits and suggestions that continue to fool the world into thinking that I have a reasonable grasp of the rules of grammar. Thanks to my sons, Eli and Abram, for making me smile every day and for the constant reminder to see the wonder and joy in all things.

Finally, thanks to my parents for their many years of love and encouragement. Without their continued support, I would not have enjoyed so many successes, and this book would not have been possible.

This book is dedicated to Anne and Frank Viscardi. Thanks for keeping the cookie jar full.

Introduction

1.1 Overview 1
1.2 Topics Addressed in This Book 3
    1.2.1 Risk-Based Monitoring 3
    1.2.2 Fraud Detection 5
    1.2.3 Snapshot Comparisons 7
1.3 The Importance of Data Standards 8
1.4 JMP Clinical 9
1.5 Clinical Trial Example: Nicardipine 13
1.6 Organization of This Book 15
References 15

1.1 Overview

Pharmaceutical development is an extremely complex affair. Once a promising compound is identified, steps are taken to optimize the chemical properties and formulation, understand the pharmacokinetics, pharmacodynamics, and safety through animal testing, then introduce the drug into humans to identify an efficacious and safe dose that can address some unmet medical need. The process involves countless tests and experiments, identifying clinicians to recruit patients into clinical trials, communicating with vendors for supplies or analysis, and routinely contacting regulatory agencies to ensure standards are met or to disclose any safety signals. There are study data to retrieve, monitor, and prepare for analysis and submission; frequent reviews to identify safety and quality concerns; scores of statistical analyses to perform; and study reports to author. This effort is made even more difficult for multinational trials: Documents require translation; differing time zones and holidays affect schedules; supplies, samples, and personnel travel long distances; and new rules and requirements are applied based on local regulatory bodies. And all of this has to happen in a timely fashion; patents are of limited duration and allow a brief opportunity for the sponsor to recoup its investment.

Given the multitude of tasks and issues just described, the numerous parties involved, and the many pitfalls that can sideline a drug (or vaccine or medical device), it is a bit surprising that the whole affair doesn't collapse on a routine basis. The success of the pharmaceutical industry is due in large part to its adherence to the processes and procedures needed for achieving various goals, as well as its commitment to detailing the resolution of problems that may occur along the way. Standard Operating Procedures (SOPs) are documents that outline the steps necessary to ensure that everything is carried out completely and accordingly, including communicating with the relevant parties both internal and external to the company. SOPs are a regular part of training for all departments and, depending on the topic, may require annual review. There are often SOPs on how to prepare and validate analyses for the final study report, how to address safety concerns that may occur at clinical sites, how to effectively incorporate and manage data monitoring committees for interim study review, and what to do should a regulatory agency perform an audit or inspection. Once in a while, you may find an SOP on SOPs!

So why include a discussion on process and procedure in a book about clinical trial software? For one, the pharmaceutical industry is at a crossroads. Productivity has dropped, clinical trials have become increasingly complex, and the costs of conducting them have skyrocketed [1–4]. Controlling these costs is essential, especially when faced with a low probability of success along the development pathway toward the marketplace. Said costs are ultimately passed on to the consumer; reining in these costs is one way to keep drugs affordable for the people who need them. According to Venet and coauthors, if costs continue to rise at the current pace, clinical trials to establish efficacy and tolerability will become impossible to conduct. This will make drugs unavailable for areas of unmet need, stifling innovation in established treatment areas, or placing an extreme price burden on consumers and health care systems [4].

Many innovations have been suggested and developed to streamline the pharmaceutical development process and improve the likelihood of clinical and regulatory success. For example, adaptive design methodologies allow for early stopping of a clinical trial in the presence of overwhelming efficacy or excess toxicity, or when the novel compound has little chance to distinguish itself from control. Extensive modeling and simulation exercises are used to suggest the most successful path forward in a clinical program based on the available data and reasonable assumptions based on past development. Patient enrichment based on genomic markers is used to select a study population more likely to receive benefit from the drug, resulting in smaller clinical trials. Some innovations have more to do with the operational aspects of clinical trials. These include electronic case report forms (eCRFs), new technologies for collecting diary data or obtaining laboratory samples, or new software that enables the efficient review of data for quality and safety purposes. And still other innovations involve the regulatory submission and review process through electronic submissions and data standards.

While pharmaceutical development is driven by process and performance, it can be slow to implement new ideas, even if they are shown to have substantial benefit. First and foremost, people are naturally resistant to change, particularly if they are comfortable in how they perform a given task. Second, the pace at a pharmaceutical company rarely slows. Individuals involved with evaluating new software, products, or techniques still have to keep the trials for which they are responsible operating smoothly while meeting or exceeding timelines. And clinical trials can last many years; finding an opportunity to implement changes may be difficult. Third, changes brought about through innovation are rarely innocuous; they can affect the processes and performance for multiple groups of individuals. This last point is particularly important to consider, since in order to implement a new method successfully, one must anticipate the growing pains and hiccups that may occur along the way.

This book is concerned with innovating the data review process for clinical trials by introducing software and techniques to steer clear of the manual "examine every data point" methods of days past. The new approach moves the reviewer from the static paper environment into an interactive and visual one. First, I promote the centralized and programmatic review of clinical trial data for signals that would indicate safety or quality problems at the clinical sites. Second, I describe methods to help uncover patient and investigator misconduct within the clinical trial. Finally, I discuss some ways to accelerate clinical trial reviews so that reviewers do not spend precious time on previously examined data. Throughout this book, I illustrate the various concepts and techniques using JMP Clinical, which I describe in "1.4 JMP Clinical" on page 9.

Any clinical trial is the result of the efforts of a diverse team. This team includes clinicians, statisticians, data managers, programmers, regulatory associates, and monitors, to name a few. Every team member has a role to play in the review of trial data. Each individual brings a unique skill set important for understanding patient safety, protocol adherence, or data insufficiencies that can affect the final analysis, clinical study report, and subsequent regulatory review and approval. For all of the aforementioned roles, aspects of JMP Clinical can streamline day-to-day work and provide new insights. It is my sincere hope that any individual from the clinical trial team can make use of this book and the examples contained within to streamline, accelerate, and enrich their reviews of clinical trial data. The few places where I get a bit technical or present SAS code or JMP scripts can be skipped by the average reader with no loss to their understanding. Major topics of this book are described in the next section. Each chapter is relatively self-contained so that the reviewer can read sections important to the task at hand.

1.2 Topics Addressed in This Book

1.2.1 Risk-Based Monitoring

Since 1990, the International Conference on Harmonisation (ICH) has brought together the regulatory bodies of the European Union, Japan, and the United States. The mission of the ICH is to define a set of technical and reporting guidelines for clinical trials to minimize the testing required in humans and animals to what is absolutely necessary to establish efficacy and safety, reduce development times, and streamline the regulatory review process. In particular, ICH Guideline E6 outlines standards for Good Clinical Practice (GCP) in the design, conduct, and reporting of clinical trials involving human participants [5]. GCP has two primary goals: to protect the well-being of subjects involved in a clinical trial and to maintain a high level of data quality to ensure the validity and integrity of the final analysis results.

Guideline E6 suggests that clinical trial data should be actively monitored to ensure data quality. Despite passages that state "the sponsor should determine the appropriate extent and nature of monitoring" and "statistically controlled sampling may be an acceptable method for selecting the data to be verified," recent practice for pharmaceutical trials has often shown a brute-force approach to source data verification (SDV) of respective CRFs through on-site monitoring [5–7]. The recent Food and Drug Administration (FDA) guidance document on risk-based monitoring (defined later) suggests a few reasons as to why this may have occurred [3]. First, the on-site monitoring model may have been (incorrectly) perceived as the preferred approach of the FDA. Second, the FDA document suggests that the agency places more emphasis on centralized monitoring than what may have been feasible at the time ICH E6 was finalized (there have been considerable technical and analytical advances in the 17 years since ICH E6 was written). While language in E6 refers to central monitoring, it does state a need for on-site monitoring "before, during, and after the trial."

However the pharmaceutical industry arrived at the current practice for clinical trial monitoring, it is now generally accepted by industry and multiple regulatory agencies that the process needs to change [1,3,8–10]. Such extensive on-site review is time consuming, expensive (up to a third of the cost of a clinical trial), and—as is true for any manual effort—limited in scope and prone to error [1,4,11–15]. In contrast to on-site monitoring, risk-based monitoring (RBM) makes use of central computerized review of clinical trial data and site metrics to determine whether clinical sites should receive more extensive quality review through on-site monitoring visits. There are many benefits to centralized review beyond cost: Statistical and graphical checks can determine the presence of outliers or unusual patterns in the data, comparisons can be made between sites to assess performance and identify potentially fraudulent data or miscalibrated or faulty equipment, and issues can be identified and resolved while the trial is ongoing.

Changing current monitoring practices to a risk-based approach will likely take time; the industry must become comfortable with a reduced presence at clinical sites and implement procedures for the remote review of clinical data, statistical sampling of data for SDV, and targeted monitoring practices. However, it is clear that the reliance on SDV, a major focus of current on-site monitoring practice, is increasingly viewed to have little to no positive impact on study conclusions [4,12]. In its position paper on RBM, TransCelerate BioPharma Inc. notes that only 7.8% of the queries generated from nine sample studies were the result of SDV, a huge investment for minimal return on data quality [1]. An example from a large international, multicenter trial found that of the issues identified, 28.4% could have been identified from the study database, and a further 66.8% could have been identified with some additional centralized edit-checks in place [16]. Further, Nielsen and coauthors illustrate that a reduced SDV monitoring approach could locate all critical queries from a pool of 30 completed clinical trials [17].

However, centralized review can only identify issues contained within the study database or other routinely collected information [4]. On-site monitoring may still be required to assess the quality of overall trial conduct, including whether appropriate regulatory documentation is available, the staff is familiar with and committed to the protocol, the staff is appropriately trained, and trial resources are adequate, well-maintained, and functional [3,5,6]. Recent literature has suggested that a diversified approach to monitoring, including centralized statistical and programmatic checks, can identify deficiencies that would otherwise go unnoticed with on-site review alone [4,6,18]. When issues are identified to a degree that may suggest a more systemic problem at the site, targeted on-site monitoring activities can be applied according to the extent of the problem and the importance of the data to the conclusions of the study [1]. The literature also stresses this important point: Data does not need to be error-free to provide reliable results from a clinical trial [1,3,6]. Finally, in addition to the risk-based methods described previously, the literature suggests a proactive approach to quality and safety through appropriate trial and CRF design, well-defined study procedures, and sufficient training of site personnel.

Chapters 2 and 3 of this book discuss an implementation of RBM within JMP Clinical that keeps to the recommendations of TransCelerate BioPharma [1]. The current application makes use of the clinical trial database and allows the team to supplement this information with any other data captured at the site level in order to assess the performance of the sites. Making use of the study data for RBM eliminates any unnecessary redundancies for similar data tracked external to the database, as well as the need for any potential reconciliation, should discrepancies arise. Chapter 2 introduces basic concepts of RBM and how to conduct reviews within JMP Clinical. Chapter 3 describes how users can customize their analysis and review experience.

1.2.2 Fraud Detection

Fraud is an important subset of topics involving data quality, one that perhaps conjures images of Sherlock Holmes (or Scooby-Doo and the Gang) on the hunt for clues to apprehend the bad guy. Quality issues in clinical trials can be due to a number of factors, among them carelessness (such as transcription errors), contamination of samples, mechanical failures, or miscalibrated equipment, poor planning (e.g., lack of appropriate backups or contingency planning should problems occur), poor training in trial procedures, and fraud. Fraud stands out among other quality issues in that there is a "deliberate attempt to deceive" or the "intention to cheat" [18]. In this book, we consider both patient- and investigator-perpetrated fraud in clinical trials. For investigators, fraud is often viewed as fabricating, manipulating, or deleting data. Examples of this behavior include deleting data highlighting a safety concern, propagating (carrying forward) data to avoid performing additional testing, or the wholesale manufacture of one or more patients at the site. For patients, enrolling at two or more clinical trials sites (usually for additional financial compensation, or access to additional drugs or medical services) is particularly problematic. This, of course, violates assumptions of statistical independence necessary for many statistical tests. In practice, subjects with multiple enrollments become an accounting and reporting nightmare for the trial team.

Despite a bevy of statistical and graphical tools available to identify unusual data, fraud is extremely difficult to diagnose. For one, many of the methods used to identify misconduct at a center involve comparisons against other clinical trial sites. Such analyses could identify natural differences in patient population or variations in technique between the sites that wouldn't constitute fraudulent behavior. Further, as we'll see later on, analyses motivated by a need to identify a particular type of malfeasance can detect data anomalies that actually have perfectly reasonable explanations. In general, stating that any unusual findings are explicitly due to fraud may require evidence beyond what's available in the clinical trial database [19].

It is believed that fraud in clinical trials is rare. Buyse and co-authors estimate the proportion of investigators engaging in misconduct below 1%, though they suggest cases may go either undiagnosed or unreported; additional cited works therein show few to no instances of fraud [18]. Two recent publications describe higher than expected rates of scholarly retractions in the life science and biomedical literature, often due to fraud or suspected fraud [20,21]. Further, in a survey of statisticians, half of 80 respondents reported awareness of fraud or deliberate deception in at least one project in the preceding 10 years, though there were some concerns about the survey's response rate and the lack of a clear definition of fraud [22,23]. Although far from a scientifically conducted poll, I obtained a similar response from an audience of approximately 35 statisticians during a section of talks on clinical trial fraud at the 2013 Statisticians in the Pharmaceutical Industry (PSI) Annual Meeting. In other surveys, misconduct was or was considered unlikely to occur [24,25].

Though the preceding paragraph paints an inconsistent picture, clinical trial fraud is likely underestimated for several aforementioned reasons. First, instances have conceivably gone undiagnosed due to a lack of available tools and training for uncovering fraud. Part of this may be due to the past overreliance on manual on-site monitoring techniques, which made comparisons across CRF pages, subjects, time, and clinical sites difficult. Further, even if unusual data are identified, going that additional step to confirm any misdoing may prove unsuccessful. Second, even if suspected fraud is detected, it may go unreported over fears that the negative publicity can severely damage the perception of an organization among regulatory agencies, patients, and the general public. Even if the sponsor has behaved entirely appropriately, such attention can bring increased scrutiny and pressure to the clinical trial and larger development program [26].

Recommendations to prevent clinical trial misconduct include simplifying study entry criteria, minimizing the amount of data collected, and ensuring sufficient and varied trial monitoring [4,6,18,26]. Even in the presence of incorrect data due to manipulation or other quality issues, trial integrity will be preserved in most cases, most often due to randomization and blinding of study medication, or because the anomalies are limited to few sites [1,3,6,18,26]. In general, clinical trial data are highly structured, and human beings are bad at fabricating realistic data, particularly in the many dimensions that would be required for it to appear plausible [4,18,19,26–30]. So this begs the question: If clinical trial fraud is so uncommon, with seemingly limited potential to seriously compromise the results of the trial, then why bother looking for it at all?

The simplest answer is that an ounce of prevention is worth a pound of cure [18]. It is much easier to identify problems as they occur while the trial is ongoing, with the opportunity to resolve the issue or modify the trial as needed. Compare this to the scenario of finding a systemic problem once the trial has been unblinded and the final study results have been prepared. At this point, there are fewer options available to the study team to find an appropriate solution (particularly when their every action will be scrutinized due to the availability of randomization codes). Defining a series of statistical and graphical checks to be implemented on a regular basis is a minimal investment for the team to make to prevent potential catastrophe.

Most important, however, we look for quality issues and misconduct because of GCP—to protect the rights and well-being of the patients enrolled in our clinical trials. Monitoring ensures that trial participants receive the best possible care and are protected from any potential wrongdoing. This is equally true for future patients who hope to use the new treatment to improve their quality of life. The rights and safety of the patient are exactly why methodologies in data quality and fraud detection should be a regular part of our statistical training. Evans states that "a perfect analysis on the wrong data can be much more dangerous than an imperfect analysis of correct data"; he suggests this is reason enough for discussing these methodologies even when others can potentially use this knowledge to avoid detection [19]. The safety and well-being of trial participants obligates us to share, collaborate, and improve methods for detecting misconduct in clinical trials. With a greater emphasis on centralized monitoring and less visibility at the clinical trial site, it is important to have a set of robust methods in place to identify fraud. Chapter 4 describes various analyses available in JMP Clinical to identify potentially fraudulent data at clinical sites. Chapter 5 presents methods to detect fraud committed by study participants.

1.2.3 Snapshot Comparisons

While Chapters 2–5 focus on assessing data quality in an efficient manner, Chapter 6 focuses on the practical considerations of review that are brought about as a result of how clinical trial data are collected and reviewed. To perform the final analysis as early as possible after the trial ends, data are collected and cleaned as they become available. Depending on the size of the trial, the number of centers, and whether enrollment is currently ongoing, new data may become available on a daily basis. This creates a constantly updating and changing database. In general, it isn't practical to update needed data sets and regenerate review reports daily—it would be difficult for reviewers to cope with this constantly moving target! Instead, an intermittent "snapshot" of the study database is taken that reflects the currently collected data and any changes since the previous snapshot. The snapshot is reviewed and necessary queries are generated to address any inconsistencies in the data or gaps in the information provided.

After a sufficient period of time, a new snapshot is generated that incorporates new data collected since the previous snapshot and any changes that were made to previously available data due to sponsor query and/or correction at the trial site. The frequency of study snapshots may depend on how much new data becomes available as well as the current lifespan of the trial. Snapshots may initially be infrequent until all study centers are operating and enrolling subjects, though snapshots occur with regularity once a sufficient number of patients are participating. Once the trial approaches last-patient-last-visit (LPLV) status, snapshots may occur very frequently, perhaps even daily, to review the final subjects' data and ensure that all needed corrections have been addressed before "locking" the trial database. Once a trial is locked, treatment codes become unblinded, and the final analysis is performed. At this point, it is expected that no further changes to the database will occur; doing so would raise suspicions that changes were made based on available treatment assignments.

As part of the review process, any number of analyses or listings may be generated to assess the quality of the data and issues pertaining to patient safety. These may include analyses to identify whether any subjects fail study eligibility criteria based on the available data, or listings to review serious adverse events or clinically significant laboratory abnormalities, dosing compliance, inconsistencies between CRF pages pertaining to visit dates or study phases, particularly noteworthy concomitant medications, and so on. During my days in the pharmaceutical industry, the biostatistics and programming group would regularly supply these listings to our clinical and regulatory team members. Inevitably, as these reviews continued, the question from our colleagues became, "Is it possible to just provide the new data so I don't have to review what I've already seen?" This seems like a perfectly reasonable request. However, the review reports were often slight modifications of the analyses that would be performed for the final clinical study report at the end of the trial. The programs rarely were written to highlight changes from one snapshot to the next, and there rarely were sufficient resources to write additional programs.

Creating informative reviews from one snapshot to the next is always more complex than just identifying and subsetting to the new data. For example, some data change during the course of the trial, so naturally there is an interest in reviewing the previous values. And to further complicate things, I found that people often wanted the new data as well as the previous data so that they could remind themselves of what led to a patient's current state of affairs, or to see if previous trends were continuing with the additional records. In other words, reviewers wanted the ability to subset or filter to new data at will. In a paper or static environment (e.g., PDF tables or listings), this request had the potential to generate twice as many analyses. Database software can track changes over time, but these tools are limited to a few individuals and are rarely available for review of the data sets that ultimately will be used for analysis and submission to regulatory agencies.

Chapter 6 of this book describes how JMP Clinical performs comparisons between current and previous data snapshots to identify new or modified values so that reviewers do not waste time on previously examined data. Further, I'll illustrate how the JMP data filter and generated review flags allow the analyst to easily switch between review summaries, including all or only newly available records. Finally, I'll examine how the extensive notes facility allows the user to create and save notes at the analysis, patient, or record level. These features provide more efficient and accurate reviews to identify and manage potential safety concerns, and to meet or exceed demanding timelines.

1.3 The Importance of Data Standards

Since its inception in 1997, the Clinical Data Interchange Standards Consortium (CDISC) has developed standards for data models, study designs, and supporting clinical trial documents. Standards for data models include the Study Data Tabulation Model (SDTM) for clinical trial data obtained from CRFs, questionnaires, and/or diaries; and the Analysis Data Model (ADaM) for derived data that support the analysis tables, listings, and figures common in drug applications [31,32]. At their core, CDISC standards are a means to streamline communications across the various parties involved in drug development, allowing for quicker review of drug submissions [33].

The SDTM model divides clinical trial data domains, a collection of logically related observations with a common topic, into a number of classes: events, findings, interventions, special purpose, and trial design. Events domains include adverse events, medical history, disposition, and protocol deviations. Examples of the findings domains are physical examinations, vital signs, ECG test results, laboratory tests, and questionnaires. Interventions domains describe concomitant medications, exposures to study treatment, and the use of substances such as alcohol, tobacco, and caffeine. Special purpose domains include demographic characteristics, general comments, and study visit attendance. Trial design domains include data describing the treatment arms, visits, and inclusion/exclusion criteria. Within each domain, variable names, labels, formats, and terms are provided, and the SDTM model states whether the inclusion of each variable in a domain is required, expected, or permissible.

The goal of ADaM is to clearly and unambiguously communicate the data behind the statistical analyses so that minimal additional manipulation is needed to generate study results (i.e., "one PROC away"). Furthermore, while SDTM typically comprises observed data separated into specific domains, ADaM can include complex derivations, such as imputation and windowing of measurements, on data spanning several domains. An additional goal of the ADaM model is traceability: providing sufficient details to allow the user to go from a tabulation to the ADaM data set to the corresponding SDTM data sets. The most important ADaM data set is ADSL, the subject-level data set. This data set provides definitions for study population flags and includes subgrouping and stratification variables, important dates such as first and last dose, and the planned and actual treatments used for each period of the study.

For analyses of clinical trial data, JMP Clinical takes advantage of CDISC standards. Perhaps more correctly, CDISC standards are a requirement! This may initially appear limiting, particularly for companies that have yet to incorporate these standards into their daily operations. However, the FDA not only provides its preferred interpretations of the standards, but it also has announced its intention to make these standards a requirement for regulatory submissions [34–36]. What are the practical benefits of requiring CDISC standards for our software? Most importantly, JMP Clinical ships with out-of-the-box functionality. JMP Clinical isn't a set of tools that requires mapping to your particular data standards, or that necessitates a team of individuals to develop and support a set of reports. Once you register your CDISC-formatted study within JMP Clinical, you're able to generate patient profiles, automate adverse event narratives, and create a host of other reports and analyses. An additional benefit of working directly from CDISC data sets is that you spend your time reviewing and analyzing the very data sets that will be submitted to the FDA or other regulatory agencies. Throughout the book, CDISC variables will be written as domain.domain-variable to be explicit. For example, USUBJID from the demography (DM) domain will be written as DM.USUBJID. When a term can apply to multiple domains, "xx" will be used to imply multiple two-letter domain codes. Interested readers can review what else SAS has available to support CDISC data standards [37].

1.4 JMP Clinical

Though I have alluded to the software, I have yet to answer the question: What exactly is JMP Clinical? Like its companion product JMP Genomics, JMP Clinical is a package that combines the analytical power of SAS with the elegant user interface and dynamic graphics of JMP. The primary purpose of this software is to simplify data discovery, analysis, and reporting for clinical trials. JMP Clinical will transform how your team currently conducts its clinical trial reviews. With its straightforward user interface and primary reliance on graphical summaries of results, everyone from the clinical trial team—including clinicians, statisticians, data managers, programmers, regulatory associates, and monitors—can explore data to identify outliers or safety or quality concerns. In this way, everyone on the team can speak the same language by using a single review tool without the hassles of importing data or results into other programs, or maintaining multiple overlapping systems.

JMP Clinical 5.0, the software version that is the focus of this book, includes JMP 11.1 and SAS 9.4 M1 for BASE SAS, SAS/GRAPH, and SAS/ACCESS to PC files, as well as SAS analytics releases 12.3 for SAS/IML, SAS/STAT, and SAS/Genetics. Also included are macros written in SAS and the JMP Scripting Language (JSL) that generate the various analyses and reports that are specific to JMP Clinical. Users familiar with JMP can expect a similar experience; analyses and reports are available from a menu-driven, point-and-click interface (Figure 1.1 on page 10). Generating reports often involves compiling a SAS program under the hood that uses the options selected from JMP dialogs to perform the various data manipulations and analyses. The results of these analyses are surfaced to the user using JSL to make them interactive and dynamic for further exploration of interesting signals. Though this product makes use of SAS, it is neither expected nor required that the user have any SAS programming experience.

Figure 1.1 JMP Clinical Starter Menu

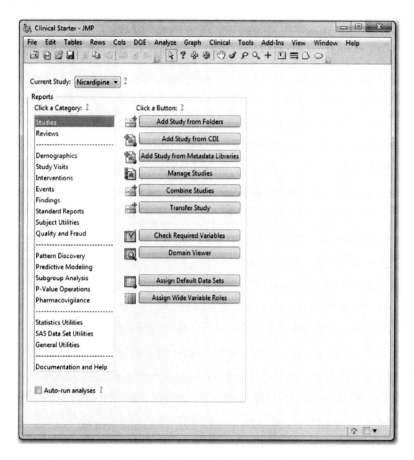

While JMP Clinical is a desktop product, it has the capability to access data and run analyses using a SAS metadata server, or access data and run analyses for studies defined using the SAS Clinical Data Integration (CDI) product. Though the review experience is generally the same, there are some important differences when operating in server mode that I will make note of throughout this book. In general, however, server profiles will be created by an experienced IT professional within your organization (see **Add SAS Server Profile** in the documentation). In addition, this individual may have the responsibility to add and manage studies as new studies or snapshots become available. When connecting to a server (**File > SAS Server Profiles > Select Profile**), studies for which you have been granted access will appear in your **Current Study** drop-down menu. All metadata (including notes and RBM files) will be shared among those individuals with access to the study. When operating locally, each user must register studies available on the local network to his or her particular instance of JMP Clinical.

Because the audience for this book is particularly diverse, many readers may be more familiar with SAS than with JMP. When I was in the pharmaceutical industry, I would have counted myself among this group of individuals; I wrote everything in SAS and had no experience whatsoever with JMP. In retrospect, this was unfortunate. The power of JMP lies in its ability to quickly and easily explore data in a graphical and interactive environment. Want a regression plot? Drag and drop two variables into the **Graph Builder** platform. See an outlier? Point and click to highlight the offending observation in the data table. It's much easier to identify and further explore anomalies from a picture than it is from a listing or table full of carefully constructed summary statistics (though JMP provides statistics as well). By the end of this book, I hope to convince other SAS users of the benefits of JMP.

Some SAS-only users may have concerns about reproducing a particular visual display when working in an interactive environment. However, every analysis or graph in JMP produces the underlying script that generates the result, and this script can be used to regenerate results at any time. For example, go to **Help > Sample Data** and open the **Big Class** data table from the alphabetical list of data files. Now go to **Analyze > Distribution** and select age, sex, height, and weight as **Y, Columns** and click **OK** (Figure 1.2 on page 11). Feel free to modify the output by closing panels of summary statistics or adding additional plots as I have done (Figure 1.3 on page 12). From the output, click the red triangle by the text Distributions, then go to **Script > Save Script to Script Window**. The new window displays the underlying script for the summary results (Figure 1.4 on page 13). If I click the **Run Script** button in the toolbar (the red running man), the results in Figure 1.3 on page 12 will be regenerated.

Figure 1.2   JMP Distribution Dialog

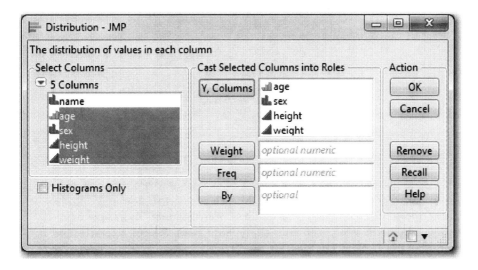

Figure 1.3 Big Class Distribution Output

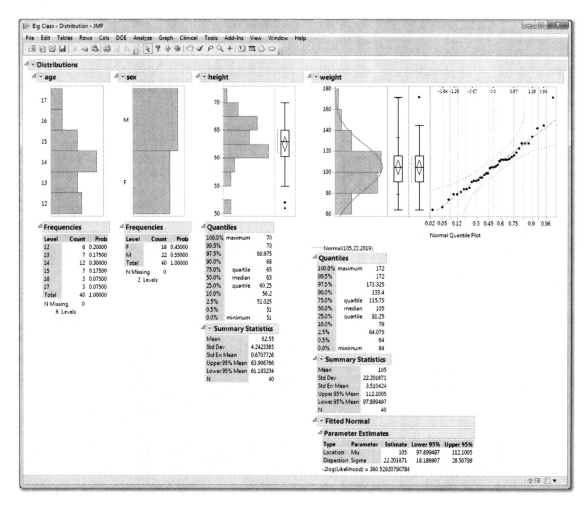

As mentioned previously, JMP Clinical makes use of CDISC variables that are required within each domain. However, in order for some reports within JMP Clinical to function appropriately or provide the greatest detail, there are often additional variable requirements. To assess what important CDISC variables may be missing from your study, run **Check Required Variables** under the **Studies** menu from the Clinical Starter. This report will summarize any analyses that may not run as a result of certain missing variables.

Users new to JMP can review *JMP Essentials, Second Edition* for help navigating the software and performing basic analyses and graphing [38]. Individuals wanting a better understanding of JMP for statistical analysis can read *JMP Start Statistics: A Guide to Statistics and Data Analysis Using JMP, Fifth Edition* [39]. *Jump into JMP Scripting* provides numerous examples and details for those users looking to master JSL [40]. If these books aren't readily available, the **Help** menu provides access to additional books, tutorials, and sample data. Assistance for JMP Clinical is available from **Help > Books > JMP Life Sciences User Guide**.

Figure 1.4 Big Class Distribution Script

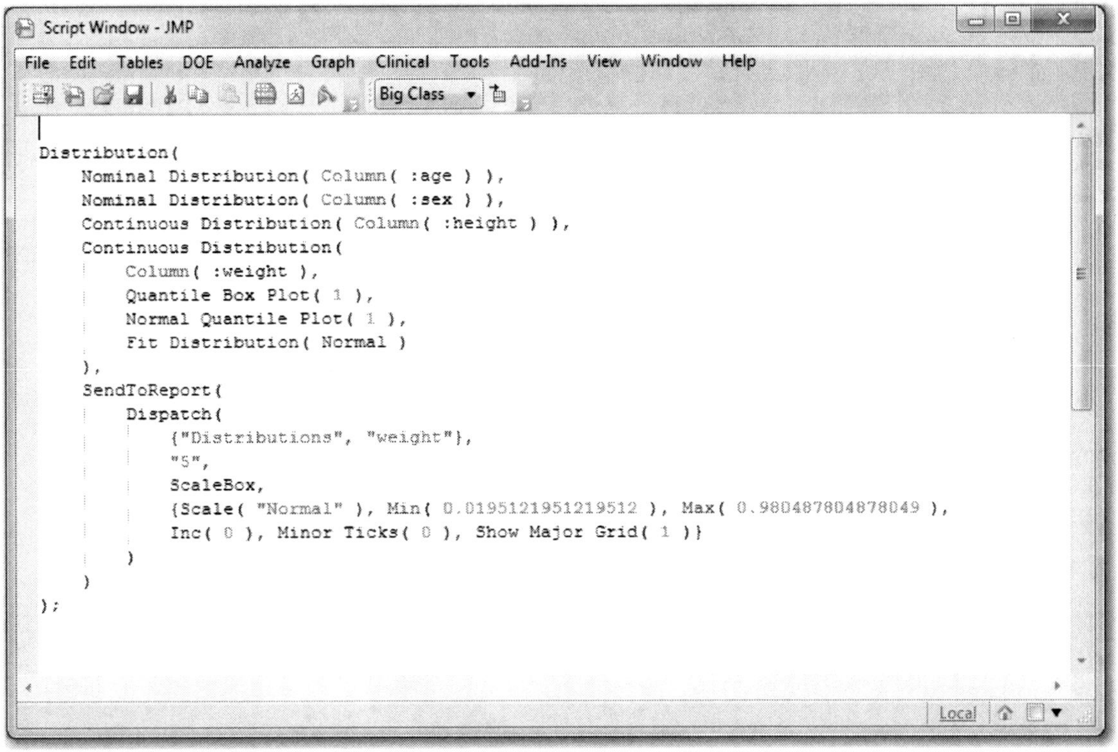
Script Window - JMP
File Edit Tables DOE Analyze Graph Clinical Tools Add-Ins View Window Help
Big Class
Distribution(
Nominal Distribution(Column(:age)),
Nominal Distribution(Column(:sex)),
Continuous Distribution(Column(:height)),
Continuous Distribution(
Column(:weight),
Quantile Box Plot(1),
Normal Quantile Plot(1),
Fit Distribution(Normal)
),
SendToReport(
Dispatch(
{"Distributions", "weight"},
"5",
ScaleBox,
{Scale("Normal"), Min(0.0195121951219512), Max(0.980487804878049),
Inc(0), Minor Ticks(0), Show Major Grid(1)}
)
)
);
Local

1.5 Clinical Trial Example: Nicardipine

JMP Clinical ships with sample data from a randomized, placebo-controlled clinical trial of nicardipine hydrochloride, a calcium channel blocker used to treat high blood pressure, angina, and congestive heart failure [41]. The drug comes in both oral and intravenous formulations. Because nicardipine also shows activity on blood vessels in the brain, this clinical trial was designed to ascertain whether there was any benefit in using nicardipine in the treatment of patients who experienced a subarachnoid hemorrhage (SAH, bleeding between the brain and the tissues that cover the brain). Patients were dosed with either intravenous nicardipine (up to 15 mg/kg/hr) or placebo for up to two weeks with the goal of reducing the incidence of delayed cerebral vasospasm, a leading cause of death for individuals experiencing an SAH. The trial randomized 906 subjects at 41 centers in the United States and Canada; 902 participants at 40 sites received treatment and constitute the Safety Population. The remaining four subjects are labeled as screen failures in the sample data. Throughout the book, nicardipine will refer to the drug, while **Nicardipine** will refer to the clinical trial registered within JMP Clinical.

When JMP Clinical first launches, the **Nicardipine** study will automatically be registered for use on the desktop using **Add Study From Folders**. Two folders of data sets are supplied, one for ADaM, in this case only the subject-level data set ADSL; and an SDTM folder containing the following domains: AE (adverse events), CM (concomitant medications), DM (demography), DS (Disposition), EG (ECG test results), EX (exposure), LB (laboratory test results), MH (medical history), SV (subject visits), and VS (vital signs). In case Nicardipine is not available on the desktop within the **Current Study** list, it can be added by going to **Studies > Add Study From Folders > Settings > Load > Nicardipine > OK > Run** (Figure 1.5 on page 14). To make **Nicardipine** available to everyone on a server, a **SAS Server Profile** first needs to be defined under the **File** menu (see **Add SAS Server Profile** in the documentation). Next, the server can be selected and **Nicardipine** can be added using settings similar to those previously described, though the data will first need to be copied to a location on the server (with separate folders for SDTM and ADaM files). In general, however, when operating on a server, a single individual will likely have the responsibility for adding and managing studies within the software. Studies that you have access to should be available in your **Current Study** drop-down list when JMP Clinical is launched. To work through the examples in this book, I recommend that you work with a local copy of **Nicardipine** to minimize any confusion.

Figure 1.5 Add Study from Folders Nicardipine Sample Setting

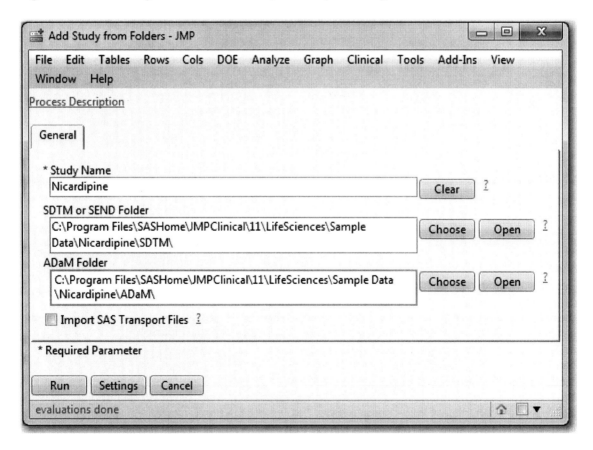

1.6 Organization of This Book

While this chapter has served as an introduction and brief literature review, the remaining chapters of this book describe analyses in more specific detail using the **Nicardipine** study to illustrate the various methodologies. Chapters 2 and 3 discuss an implementation of RBM based on the recommendations of TransCelerate BioPharma [1]. Here, an artificial example is created for Nicardipine using simulated data for several site metrics typically not found within the clinical trial database. Locations of trial sites were modified to include countries in Europe and Asia, and random cities were selected to represent site locations. Chapters 4 and 5 present analyses to identify potentially fraudulent data in your clinical trial. Chapter 6 describes snapshot comparison tools to highlight new and modified data and briefly summarizes the notes facilities of JMP Clinical. Finally, Chapter 7 serves as a brief epilogue. At the conclusion of each chapter, I suggest areas for possible future development. Your feedback is important to prioritize these potential improvements for future versions of the software.

References

1. TransCelerate BioPharma Inc. (2013). Position paper: Risk-based monitoring methodology. Available at: http://transceleratebiopharmainc.com/.

2. Scannell JW, Blanckley A, Boldon H & Warrington B. (2012). Diagnosing the decline in pharmaceutical R&D efficiency. *Nature Reviews Drug Discovery* 11: 191–200.

3. US Food & Drug Administration. (2013). Guidance for industry: Oversight of clinical investigations—a risk-based approach to monitoring. Available at: http://www.fda.gov/downloads/Drugs/.../Guidances/UCM269919.pdf.

4. Venet D, Doffagne E, Burzykowski T, Beckers F, Tellier Y, Genevois-Marlin E, Becker U, Bee V, Wilson V, Legrand C & Buyse M. (2012). A statistical approach to central monitoring of data quality in clinical trials. *Clinical Trials* 9: 705–713.

5. International Conference of Harmonisation. (1996). E6: Guideline for Good Clinical Practice. Available at: http://www.ich.org/fileadmin/Public_Web_Site/ICH_Products/Guidelines/Efficacy/E6_R1/Step4/E6_R1__Guideline.pdf.

6. Baigent C, Harrell FE, Buyse M, Emberson JR & Altman DG. (2008). Ensuring trial validity by data quality assurance and diversification of monitoring methods. *Clinical Trials* 5: 49–55.

7. Morrison BW, Cochran CJ, White JG, Harley J, Kleppinger CF, Liu A, Mitchel JT, Nickerson DF, Zacharias CR, Kramer JM, Neaton JD. (2011). Monitoring the quality of conduct of clinical trials: a survey of current practices. *Clinical Trials* 8: 342–349.

8   Clinical Trials Transformation Initiative. Effective and efficient monitoring as a component of quality assurance in the conduct of clinical trials. Available at: http://www.ctti-clinicaltrials.org/what-we-do/study-conduct/monitoring. Accessed 23May2014.

9   European Medicines Agency. (2011). Reflection paper on risk based quality management in clinical trials. Available at: http://www.ema.europa.eu/docs/en_GB/document_library/Scientific_guideline/2013/11/WC500155491.pdf.

10 Medicines and Healthcare Products Regulatory Agency. (2011). Risk-adapted approaches to the management of clinical trials of investigational medicinal products. Available at: http://www.mhra.gov.uk/home/groups/l-ctu/documents/websiteresources/con111784.pdf.

11 Eisenstein EL, Lemons PW, Tardiff BE, Schulman KA, Jolly MK & Califf RM. (2005). Reducing the costs of phase III cardiovascular clinical trials. *American Heart Journal* 149: 482–488.

12 Tantsyura V, Grimes I, Mitchel J, Fendt K, Sirichenko S, Waters J, Crowe J & Tardiff B. (2010). Risk-based source data verification approaches: pros and cons. *Drug Information Journal* 44: 745–756.

13 Funning S, Grahnén A, Eriksson K & Kettis-Linblad A. (2009). Quality assurance within the scope of good clinical practice (GCP): what is the cost of GCP-related activities? A survey with the Swedish association of the pharmaceutical industry (LIF)'s members. *The Quality Assurance Journal* 12: 3–7.

14 Grieve AP. (2012). Source data verification by statistical sampling: issues in implementation. *Drug Information Journal* 46: 368–377.

15 Usher RW. (2010). PhRMA bioresearch monitoring committee perspective on acceptable approaches for clinical trial monitoring. *Drug Information Journal* 44: 477–483.

16 Bakobaki JM, Rauchenberger M, Joffe N, McCormack S, Stenning S & Meredith S. (2012). The potential for central monitoring techniques to replace on-site monitoring: findings from an international multi centre clinical trial. *Clinical Trials* 9: 257–264.

17 Nielsen E, Hyder D & Deng C. (2014). A data-driven approach to risk-based source data verification. *Therapeutic Innovation and Regulatory Science* 48: 173–180.

18 Buyse M, George SL, Evans S, Geller NL, Ranstam J, Scherrer B, LeSaffre E, Murray G, Elder L, Hutton J, Colton T, Lachenbruch P & Verma BL. (1999) The role of biostatistics in the prevention, detection and treatment of fraud in clinical trials. *Statistics in Medicine* 18: 3435–3451.

19 Evans S. (2001). Statistical aspects of the detection of fraud. In: Lock S & Wells F, eds. *Fraud and Misconduct in Biomedical Research, Third Edition*. London: BMJ Books.

20 Fang FC, Steen RG & Casadevall A. (2012). Misconduct accounts for the majority of retracted scientific publications. *Proceedings from the National Academy of Sciences* 109: 17028–17033.

21 Grieneisen ML & Zhang M. (2012). A comprehensive survey of retracted articles from the scholarly literature. *Public Library of Science (PLoS) ONE* 7(10): 1–15.

22 Ranstam J, Buyse M, George SL, Evans S, Geller NL, Scherrer B, Lesaffre E, Murray G, Edler L, Hutton JL, Colton T & Lachenbruch P. (2000). Fraud in medical research: An international survey of biostatisticians. *Controlled Clinical Trials* 21: 415–427.

23 Ellenberg S. (2000). Fraud is bad, studying fraud is hard. *Controlled Clinical Trials* 21: 498–500.

24 Al-Marzouki SA, Roberts I, Marshall T & Evans S. (2005). The effect of scientific misconduct on the results of clinical trials: A Delphi survey. *Contemporary Clinical Trials* 26: 331–337.

25 Baerlocher MO, O'Brien J, Newton M, Gautam T & Noble J. (2010). Data integrity, reliability and fraud in medical research. *European Journal of Internal Medicine* 21: 40–45.

26 Weir C. & Murray G. (2011). Fraud in clinical trials: detecting it and preventing it. *Significance* 8: 164–168.

27 Akhtar-Danesh A & Dehghan-Kooshkghazi M. (2003). How does correlation structure differ between real and fabricated data-sets? *BMC Medical Research Methodology* 3: 1–9.

28 Al-Marzouki SA, Evans S, Marshall T & Roberts I. (2005). Are these data real? Statistical methods for the detection of data fabrication in clinical trials. *British Medical Journal* 331: 267–270.

29 Taylor RN, McEntegart DJ & Stillman EC. (2002). Statistical techniques to detect fraud and other data irregularities in clinical questionnaire data. *Drug Information Journal* 36: 115–125.

30 Wu X & Carlsson M. (2011). Detecting data fabrication in clinical trials from cluster analysis perspective. *Pharmaceutical Statistics* 10: 257–264.

31 CDISC Submission Data Standards Team. (2013). Study data Tabulation Model Implementation Guide: Human Clinical Trials, Version 3.2. Round Rock, TX: Clinical Data Interchange Standards Consortium.

32 CDISC Analysis Data Model Team. (2009). Analysis Data Model (ADaM), Version 2.1. Round Rock, TX: Clinical Data Interchange Standards Consortium. (UPDATE)

33 Zink RC & Mann G. (2012). On the importance of a single data standard. *Drug Information Journal* 46: 362–367.

34 US Food and Drug Administration, Center for Drug Evaluation and Research. (2011). CDER Common Data Standards Issues Document, Version 1.1. http://www.fda.gov/downloads/Drugs/DevelopmentApprovalProcess/FormsSubmissionRequirements/ElectronicSubmissions/UCM254113.pdf.

35 US Food and Drug Administration. (2012). PDUFA reauthorization performance goals and procedures fiscal years 2013 through 2017. http://www.fda.gov/downloads/ForIndustry/UserFees/PrescriptionDrugUserFee/UCM270412.pdf.

36 US Food and Drug Administration. (2013). Position Statement for Study Data Standards for Regulatory Submissions. http://www.fda.gov/ForIndustry/DataStandards/StudyDataStandards/ucm368613.htm.

37 Holland C & Shostak J. (2012). *Implementing CDISC Using SAS: An End-to-End Guide*. Cary, NC: SAS Institute Inc.

38 Hinrichs C & Boiler C. (2014). *JMP Essentials, Second Edition*. Cary, NC: SAS Institute Inc.

39 Sall J, Lehman A, Stephens M & Creighton L. (2012). *JMP Start Statistics: A Guide to Statistics and Data Analysis Using JMP, Fifth Edition*. Cary, NC: SAS Institute Inc.

40 Murphrey W & Lucas R. (2009). *Jump into JMP Scripting*. Cary, NC: SAS Institute Inc.

41 Haley EC, Kassell NF & Torner JC. (1993). A randomized controlled trial of high-dose intravenous nicardipine in aneurysmal subarachnoid hemorrhage. *Journal of Neurosurgery* 78: 537–547.

Risk-Based Monitoring: Basic Concepts

2.1 *Introduction* 19
2.2 *Risk Indicators* 22
    2.2.1 Individual Risk Indicators 22
    2.2.2 Overall Risk Indicators 42
    2.2.3 Default Risk Thresholds 44
    2.2.4 Default Actions 48
2.3 *Geocoding Sites* 49
2.4 *Reviewing Risk Indicators* 53
    2.4.1 Site-Level Risk 53
    2.4.2 Country-Level Risk 66
    2.4.3 Subject-Level Risk 70
2.5 *Final Thoughts* 73
References 73
Appendix 75
    Walk-through of the RBM Reports 75
    Definitions of Risk Indicators and Important Terms in Pseudo-code 76

2.1 Introduction

Recent interest in risk-based monitoring (RBM) is driven primarily by the increasingly unsustainable costs of conducting clinical trials [1,2]. But what is driving these expenses?

Eisenstein et al. report that 25–30% of the costs of two hypothetical cardiovascular clinical trials involve site-management activities [3]. Tantsyura and coauthors report that up to 30% of a trial's budget can be consumed by site monitoring [4]. A Swedish survey estimates that 50% of a trial's costs are attributable to efforts related to International Conference of Harmonisation (ICH) Good Clinical Practice (GCP) activities, half of which are consumed by source data verification (SDV) [5,6,7]. Given that SDV is increasingly viewed to have little to no positive impact on study conclusions, that a majority of issues can be identified and addressed remotely, and the increasing acceptance that study databases need not be error-free to provide valid conclusions, addressing costs related to on-site monitoring activities can provide very real savings for trial sponsors [1–3,8–11].

RBM makes use of central computerized review of clinical trial data and site metrics to determine if clinical sites should receive more extensive quality review through on-site monitoring visits. While there are certainly benefits to RBM in reducing trial costs, recent regulatory guidance suggests that risk-based approaches may actually improve overall study quality by focusing the sponsor's efforts on those data and processes most important to patient safety and the integrity of the final results. Further, RBM emphasizes the need for the sponsor to be proactive by detailing how data will be reviewed, defining thresholds for risk, and outlining the processes put in place to prevent mishap as well as the necessary steps to resolve issues should they occur. In short, quality is built into the trial, ensuring that it becomes less of a reactionary brute-force process.

First and foremost, sponsors need to prospectively identify the areas that are critical to the success of the clinical trial and the safety of its participants. These areas will often include primary and secondary efficacy endpoints; appropriate storage, administration, and compliance of study medication; frequency of important safety issues such as serious adverse events (SAEs) and deaths; subject eligibility criteria; protocol deviations and study discontinuations; informed consent; and randomization, blinding, and event adjudication [1,8,12,13]. Other issues relevant to the operational aspects of the trial include timely completion of case report forms (CRFs), response to queries, and recruitment [1]. The sponsor should specify the extent to which these data will be reviewed, detailing the necessary steps to address and mitigate risk should problems arise. Many of these details can be decided in a program-level monitoring plan, such as the Integrated Quality Risk Management Plan (IQRMP) suggested by TransCelerate BioPharma Inc. However, some notable differences may arise between an individual trial and the IQRMP, depending on the study's characteristics. For example, first-in-human studies, particularly complex trial designs with one or more adaptations, or trials involving special patient populations (e.g., pediatrics), or trials with new or unfamiliar equipment or investigators may require additional levels of scrutiny of the collected data or review of additional information.

Once important data are identified, risk indicators are developed and tracked to monitor progress and performance at clinical sites, and to assess whether more formal review or intervention is warranted. For example, data for adverse events are collected that describe the ongoing safety and well-being of study participants. One possible risk indicator may average the total number of adverse events experienced by patients at a clinical site by the number of subjects randomized. An "overly large" event-per-subject value compared to the overall average across all sites may indicate a safety problem at the site. On the other hand, an "overly small" value may point to a site that is underreporting adverse events. Either issue is of potential concern, and whether the site is contacted with a routine phone call or visited in person depends on how extreme the finding is compared to other sites, the number of other related risk indicators that display excess risk, the number of subjects enrolled at the site (perhaps the current estimate is an anomaly due to small sample size), or whether this is repeat behavior for this site within the current trial or possibly from other studies. Important for any risk indicator is to predefine the direction(s) of concern, the magnitudes that would exceed important thresholds of risk, and the sponsor's action should one or more thresholds be crossed. Section 8.1.3 of the TransCelerate position paper contains a very thorough list of possible risk indicators [1].

The RBM functionality of JMP Clinical was developed using the recommendations of TransCelerate BioPharma and is available from the starter menu under **Risk-Based Monitoring** (Figure 2.1 on page 22); there are currently five report buttons available. First, **Risk-Based Monitoring** generates the primary RBM analysis using data from the Study Data Tabulation Model (SDTM) and/or Analysis Data Model (ADaM) study folders as well as any user-supplied data that is external to the trial database. Here, RBM reviews are based on a traffic-light system to indicate elevated risk, with histograms and box plots to highlight any outliers, and maps to help uncover any possible geographic trends in risk based on the locations of the clinical sites. To supplement this presentation, I'll describe some more formal statistical techniques to review risk indicators in the next chapter. **Update Study Risk Data Set** allows the user to supplement the study database with data from other sources. Information regarding queries and CRF pages taken from database management systems (DBMS), summary values from statistical programs written to determine protocol deviations, or other data taken from JMP Clinical can be entered at the site level and analyzed using **Risk-Based Monitoring**. Users can supply additional geographic information for sites within **Update Study Risk Data Set** so that site locations can be geocoded on global and country maps in **Risk-Based Monitoring**. **Map Geocoding Help** can assist with appropriate spellings or combinations of geographic information to get sites properly geocoded.

Define Risk Threshold Data Set allows the user to define data sets that apply sets of risk thresholds to the individual risk indicators created within **Risk-Based Monitoring**. In addition, users can specify how individual risk indicators should be combined to create one or more overall indicators of risk. Multiple risk threshold data sets can be defined addressing varying levels of risk; the sensitivity of the RBM analysis can be assessed by applying different risk threshold data sets. Initially, we'll make use of the **Default Risk Threshold** data set that is supplied with JMP Clinical. I'll describe how to create new risk threshold data sets toward the end of the chapter. **Manage Risk Threshold Data Sets** allows the user to delete unwanted risk threshold data sets.

Figure 2.1 Risk-Based Monitoring Menu

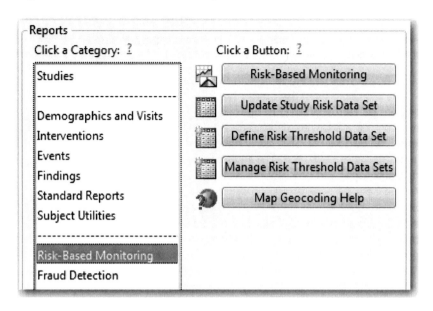

In "2.2 Risk Indicators" on page 22, I describe assumptions and variables that are used to define risk indicators that are generated automatically by JMP Clinical from the study SDTM and/or ADaM folders. I also discuss how users can supplement the study database with additional data to create their own risk indicators and define how overall indicators of risk are generated. In "2.3 Geocoding Sites" on page 49, I describe how users can easily geocode their clinical sites for a more informative analysis. "2.4 Reviewing Risk Indicators" on page 53 focuses on RBM analysis and review at site and country levels. Further, I demonstrate how to identify and review study participants from problematic sites or countries. The "Appendix" on page 75 contains "Walk-through of the RBM Reports" on page 75 needed to perform RBM analyses employing the most up-to-date data, and presents the "Definitions of Risk Indicators and Important Terms in Pseudo-code" on page 76 of RBM terminology utilized in JMP Clinical. In Chapter 3, I describe how to modify default risk thresholds and provide tools for the analyst to take their RBM analyses and reviews further.

2.2 Risk Indicators

2.2.1 Individual Risk Indicators

2.2.1.1 Risk Indicators from CDISC-Formatted Data Sets

In the very near future, CDISC-formatted data sets will be required for regulatory submissions to the U.S. Food and Drug Administration [14,15]. While it is entirely possible for a company to maintain its pre-CDISC standards and map its data to SDTM prior to submission, this is a shortsighted strategy. Most important, there are very real concerns that there could be differences between analyses based on sponsor- versus CDISC-standard data sets due to difficulties in mapping between the two. Depending on the severity of the problem, this may cause delays in the regulatory submission or review. It is much more straightforward and efficient to align systems and software to work with CDISC standards throughout the development process. This, in turn, fosters experience in the standards for the development team, and enables better communication throughout the sponsor organization and across regulators, contract research organizations (CROs), and other vendors [16].

Here, JMP Clinical makes use of the CDISC-formatted study database to generate risk indicators for RBM analysis and review. It is important to take advantage of the database as much as possible to minimize redundancy and avoid unnecessary reconciliation with similar data that are captured external to the database in a less stringent fashion. When needed data are not available from the database to define important risk indicators, we will supplement as necessary from other sources, using data from a DBMS, output from statistical programs, or output from other analyses available in JMP Clinical ("2.2.1.2 Risk Indicators from Other Data Sources" on page 31).

Figure 2.2 Risk-Based Monitoring Dialog

JMP Clinical has minimal requirements to perform analyses related to RBM, and these analyses are available from the **Risk-Based Monitoring** report. The dialog (Figure 2.2 on page 24) is very straightforward, asking the user to select which study to analyze, the set of risk thresholds to apply through the choice of **Risk Threshold Data Set**, and whether metrics based on a planned target rate of enrollment are desired in the resulting output. Also available are a number of **WHERE** statements that allow the analyst to subset to those findings or events that may be of greater interest for the current analysis. These **WHERE** statements are discussed later under the appropriate CDISC domains. For the remainder of this section, we'll discuss the requirements of the RBM analysis in a manner accessible to a majority of users; individuals interested in viewing the requirements can go to "Definitions of Risk Indicators and Important Terms in Pseudo-code" on page 76.

JMP Clinical requires the following information for analysis:

1. Demography (DM): Study Site Identifier and Country;

2. Disposition (DS): Standardized Disposition Term to determine whether a subject was a screen failure, was randomized, completed, or discontinued the trial. Here, the **WHERE** statement is used to more specifically select the records (ideally, one per subject) that indicate either subject completion or discontinuation status for the study. Details on how these records are selected by default are left to the Appendix;

3. Subject Visits (SV) or one or more CDISC Findings domains to calculate the number of weeks a patient has been on-study.

Though not required, information from the Exposure (EX) domain may be used to determine whether subjects were treated with study therapy.

If available, data from the following domains are used to generate additional metrics and risk indicators:

1. Adverse Events (AE) for risk indicators involving AEs and SAEs and to determine if any events were fatal,

2. Protocol Deviations (DV) for risk indicators illustrating poor adherence to the study protocol,

3. Inclusion/Exclusion Criterion Not Met (IE) for risk indicators of failed eligibility criteria.

Note: The **Nicardipine** trial does not have the DV or IE domains available.

The above three domains use **WHERE** statements to subset to data of interest, and can be used to assess the sensitivity of the RBM analysis. For example, if I am currently working on an ophthalmology trial, I might choose to limit the analysis to ocular AEs by specifying AECAT = "OCULAR". Alternatively, we may be concerned that investigators are underreporting the severity of events; users may view how the results of safety risk indicators change if the analysis is subset to AEs that are moderate or severe (e.g., AESEV in ('MODERATE','SEVERE')). The **WHERE** statements are very powerful in that they can be used to select findings or events based on any number of criteria. However, users need to know the specific contents (variables and values) of their domains. Variables and values can be identified using **Studies > Domain Viewer**, or interested individuals can get assistance from colleagues more familiar with the details of the study data sets. Once **WHERE** statements are defined, settings can be saved using **Settings > Save** from the **Risk-Based Monitoring** dialog. Note that while **WHERE** statements follow SAS syntax, they can usually be generated by anyone with a little practice.

Risk indicators derived from CDISC-formatted data sets are organized under the following group headings in the risk indicator data tables: Enrollment Metrics, Disposition, and Adverse Events. These variable groupings allow the user to easily highlight terms for review using the **Columns** menu in the left panel next to the data table, or select groups of terms for further analysis (Figure 2.3 on page 26). Note that this left panel can be hidden by clicking on the grey triangle in the upper left corner of the data table. Furthermore, rows or columns can be de-selected by clicking anywhere in the lower or upper triangles in the upper-left corner of the data table. Though we generally refer to all variables in the subject- and country-level risk tables as "risk indicators," only variables for which thresholds have been defined will be colored in green, yellow, or red, corresponding to mild, moderate, or severe risk.

Figure 2.3 Selecting Indicators in the Site-Level Data Table

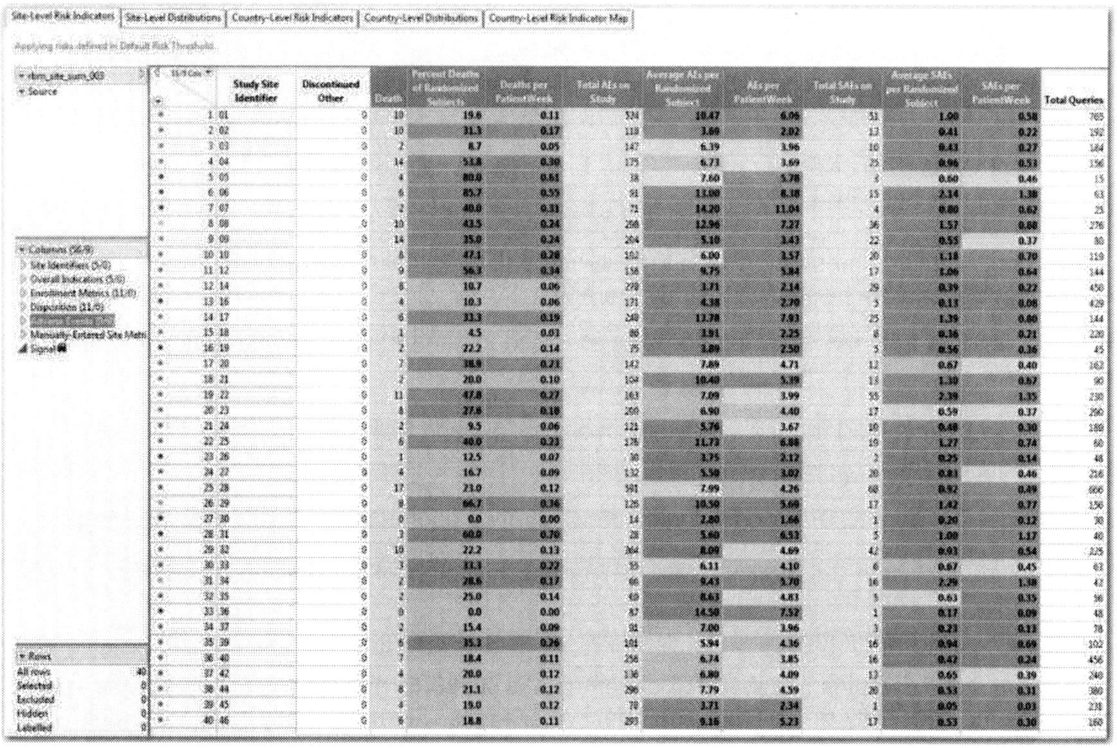

Because the RBM analysis works directly with the study database, needed data are obtained at the subject level. Individual metrics and information needed to define risk indicators are initially derived at the subject level and later summarized at the site and country levels for review in the risk indicator data tables within **Risk-Based Monitoring**. While site- and country-level summary information is presented front and center, subject-level information contributing to these summary indicators is available for review from the **Show Subjects** drill down. Further, several edit checks that examine the consistency of subject-level data are available from **Open Edit Checks** drill down. As part of the RBM exercise, comparisons are performed to help clean the study database and to ensure that the data going into the analysis are of high quality. Also, tying the RBM analysis and review back to the individual patients makes it easy to review participants at problematic sites. For example, if I identify a site that appears to have excess safety risk, I can explore these subjects in further detail by generating **Patient Profiles** or **AE Narratives**. I'll illustrate the use of the aforementioned drill downs in "2.4 Reviewing Risk Indicators" on page 53.

In addition to displaying the frequency of many important safety and quality metrics, such as the **Total AEs on Study**, JMP Clinical calculates averages for many of the more important variables in two ways (Figure 2.4 on page 27 uses **Cols > Reorder Columns**):

1 Averaged by the number of randomized subjects. For example, **Average AEs per Randomized Subject** divides the **Total AEs on Study** by **Randomized**, the total number of randomized subjects at the study site.

2 Averaged by the number of patient weeks. For example, **AEs per PatientWeek** divides the **Total AEs on Study** by **PatientWeeks on Study**, the total time subjects have been enrolled in the clinical trial.

Figure 2.4 Risk Indicators for Adverse Events

	Study Site Identifier	Country	Randomized	PatientWeeks on Study	Total AEs on Study	Average AEs per Randomized Subject	AEs per PatientWeek	Total SAEs on Study	Average SAEs per Randomized Subject	SAEs per PatientWeek
1	01	USA	51	88.14	534	10.47	6.06	51	1.00	0.58
2	02	USA	32	58.29	118	3.69	2.02	13	0.41	0.22
3	03	USA	23	37.14	147	6.39	3.96	10	0.43	0.27
4	04	FRA	26	47.43	175	6.73	3.69	25	0.96	0.53
5	05	ITA	5	6.57	38	7.60	5.78	3	0.60	0.46
6	06	USA	7	10.86	91	13.00	8.38	15	2.14	1.38
7	07	CHN	5	6.43	71	14.20	11.04	4	0.80	0.62
8	08	GBR	23	41.00	298	12.96	7.27	36	1.57	0.88
9	09	CAN	40	59.43	204	5.10	3.43	22	0.55	0.37
10	10	USA	17	28.57	102	6.00	3.57	20	1.18	0.70
11	12	DEU	16	26.71	156	9.75	5.84	17	1.06	0.64
12	14	CAN	75	129.86	278	3.71	2.14	29	0.39	0.22
13	16	USA	39	63.43	171	4.38	2.70	5	0.13	0.08
14	17	USA	18	31.29	248	13.78	7.93	25	1.39	0.80
15	18	JPN	22	38.29	86	3.91	2.25	8	0.36	0.21
16	19	CAN	9	14.00	35	3.89	2.50	5	0.56	0.36
17	20	FRA	18	30.14	142	7.89	4.71	12	0.67	0.40
18	21	ESP	10	19.29	104	10.40	5.39	13	1.30	0.67
19	22	CAN	23	40.86	163	7.09	3.99	55	2.39	1.35
20	23	USA	29	45.43	200	6.90	4.40	17	0.59	0.37
21	24	CHN	21	33.00	121	5.76	3.67	10	0.48	0.30
22	25	USA	15	25.57	176	11.73	6.88	19	1.27	0.74
23	26	DEU	8	14.14	30	3.75	2.12	2	0.25	0.14
24	27	CHE	24	43.71	132	5.50	3.02	20	0.83	0.46
25	28	CAN	74	138.86	591	7.99	4.26	68	0.92	0.49
26	29	USA	12	22.14	126	10.50	5.69	17	1.42	0.77
27	30	USA	5	8.43	14	2.80	1.66	1	0.20	0.12
28	31	ITA	5	4.29	28	5.60	6.53	5	1.00	1.17
29	32	USA	45	77.57	364	8.09	4.69	42	0.93	0.54
30	33	USA	9	13.43	55	6.11	4.10	6	0.67	0.45
31	34	CHN	7	11.57	66	9.43	5.70	16	2.29	1.38
32	35	USA	8	14.29	69	8.63	4.83	5	0.63	0.35
33	36	USA	6	11.57	87	14.50	7.52	1	0.17	0.09
34	37	ESP	13	23.00	91	7.00	3.96	3	0.23	0.13
35	39	USA	17	23.14	101	5.94	4.36	16	0.94	0.69
36	40	GBR	38	66.57	256	6.74	3.85	16	0.42	0.24
37	42	GBR	20	33.29	136	6.80	4.09	13	0.65	0.39
38	44	USA	38	64.43	296	7.79	4.59	20	0.53	0.31
39	45	JPN	21	33.29	78	3.71	2.34	1	0.05	0.03
40	46	USA	32	56.00	293	9.16	5.23	17	0.53	0.30

Unlike averaging by randomized subjects, averages based on patient weeks account for how long subjects have participated in the trial and their potential exposures to study medication. Patient weeks can also be considered a measure of "site experience." For example, imagine two sites, each with the same number of randomized subjects and queries, but subjects at Site B have been in the trial twice as long as subjects in Site A. Averages based on the number of randomized subjects will not show any difference in performance between the two sites. However, Site A is showing twice the number of issues when accounting for how long subjects have been on trial. Both ways of normalizing the data are important to understand site performance, though other definitions are certainly possible (e.g., normalizing by the number of weeks a site has been active, **Weeks Active**). Given that clinical sites enter the study and enroll subjects at different times, it's important to consider this information together, particularly if there are differences in interpretation between the two types of metrics in terms of their risk thresholds or outlier status.

Finally, Table 2.1 on page 28 summarizes the site identifiers and risk indicators derived from CDISC-formatted data sets. If a new data snapshot becomes available, go to **Studies > Manage Studies > Update Study Data and Metadata** and provide the new **SDTM** and/or **ADaM Folders**. Then the RBM analysis can be repeated on the most up-to-date study database available. Risk indicators derived from data external to the study database are entered through **Update Study Risk Data Set** and will be described in the next section. These data are updated separately from the database, and changes to these data can be made at any time.

Table 2.1 Site Identifiers and Risk Indicators from Study Database

Grouping	Variable	Details
Site Identifiers	Study Site Identifier	DM.SITEID
	Country	DM.COUNTRY
Enrollment Metrics	Total Subjects	Number of unique subjects identified, should equal randomized plus screen failures
	Randomized	Number of randomized subjects
	Screen Failure	Number of screen failures
	Percent Screen Fail of Total Subjects	Screen Failure / Total Subjects
	Treated	Number of treated subjects
	PatientWeeks on Study	Sum of the weeks subjects have been in the study

Grouping	Variable	Details
	Weeks Active	How long site has been active. Computed from Site Active Date supplied in **Update Study Risk Data Set** and maximum available date from SV or Findings domains. Remaining enrollment metrics are dependent on Site Active Date being supplied
	Randomized per Week Active	Average number of randomized subjects per week
	Expected Randomized	Number of randomized subjects expected based on the total randomized and sum of Weeks Active across all sites
	Observed Minus Expected Randomized	Randomized per Week Active minus Expected Randomized. Negative or positive numbers show under- or overperformance, respectively
	Target Randomized	Based on target rate supplied from the dialog
	Observed Minus Target Randomized	Randomized per Week Active minus Target Randomized. Negative or positive numbers show under- or overperformance from target, respectively
Disposition	Completed	Number of subjects completing the trial
	Percent Completed of Randomized Subjects	Completed / Randomized
	Percent Ongoing of Randomized Subjects	Ongoing / Randomized
	Discontinued	Number of subjects that discontinued the trial early

Grouping	Variable	Details
	Ongoing	Randomized minus Completed minus Discontinued
	Percent Discontinued of Randomized Subjects	Discontinued / Randomized
	Discontinued Due to Death	Number of subjects discontinuing due to death
	Discontinued Due to AE	Number of subjects discontinuing due to AE
	Lost to Follow Up	Number of subjects lost to follow-up
	Patient Withdrew from Study	Number of subjects that withdrew consent
	Discontinued Other	Number of subjects discontinuing not captured by other reasons
	Total Protocol Deviations	Number of protocol deviations experienced by randomized subjects at the site
	Average Deviations per Randomized Subject	Total Protocol Deviations/ Randomized
	Total Inclusion or Exclusion Not Met	Number of inclusion exclusion criteria not met experienced by randomized subjects at the site
	Average Inclusion or Exclusion Not Met per Randomized Subject	Total Inclusion or Exclusion Not Met / Randomized
Adverse Events	Death	Number of subjects that died
	Percent Deaths of Randomized Subjects	Death / Randomized
	Deaths per PatientWeek	Death / PatientWeeks on Study

Grouping	Variable	Details
	Total AEs on Study	Number of AEs experienced by subjects at the site
	Average AEs per Randomized Subject	Total AEs on Study / Randomized
	AEs per PatientWeek	Total AEs on Study / PatientWeeks on Study
	Total SAEs on Study	Number of SAEs experienced by subjects at the site
	Average SAEs per Randomized Subject	Total SAEs on Study / Randomized
	SAEs per PatientWeek	Total SAEs on Study / PatientWeeks on Study

2.2.1.2 Risk Indicators from Other Data Sources

While the study database captures a great deal of information to assess the safety and quality of the ongoing trial, it is by no means complete. For example, many data management activities track the performance of sites in terms of timely completion of CRFs and response to queries, as well as the number of queries generated or missing pages from the CRF. Study eligibility criteria collected from the CRF and reported in the IE domain are based on the assessment of the investigators at the clinical site. Often, the biostatistics and programming team will confirm these criteria using the study data, with some discrepancies to the site evaluation arising. In a similar manner, protocol deviations derived by the biostatistics team are not presented in the DV domain, though they may uncover serious deficiencies in site or patient adherence to the protocol that were not previously reported. Finally, monitoring activities may contribute additional data, such as the temperature at which the study drug is stored, discrepancies in drug dispensation compared to Interactive Voice Response System (IVRS) logs, or the number of missing regulatory documents that are to be maintained at the site. **Update Study Risk Data Set** defines a data set that captures these "manually entered" values, as well as additional geographic and other site identifier information, in order to supplement the analysis performed in **Risk-Based Monitoring**. In general, it is a good practice to create this supplemental data set prior to any RBM analysis, if only to provide geographic and site identifier data. Geographic data will be discussed in "2.3 Geocoding Sites" on page 49.

Nicardipine has a specially built example for RBM to highlight many of the analysis and review features available (Figure 2.5 on page 33). However, first time users generate this data table for their own studies; it should be blank except for the sites and respective countries currently identified in the study database (Figure 2.6 on page 34). The dialog for **Update Study Risk Data Set** is straightforward, allowing the user to select only the study.

Figure 2.5 Study Risk Data Set for Nicardipine

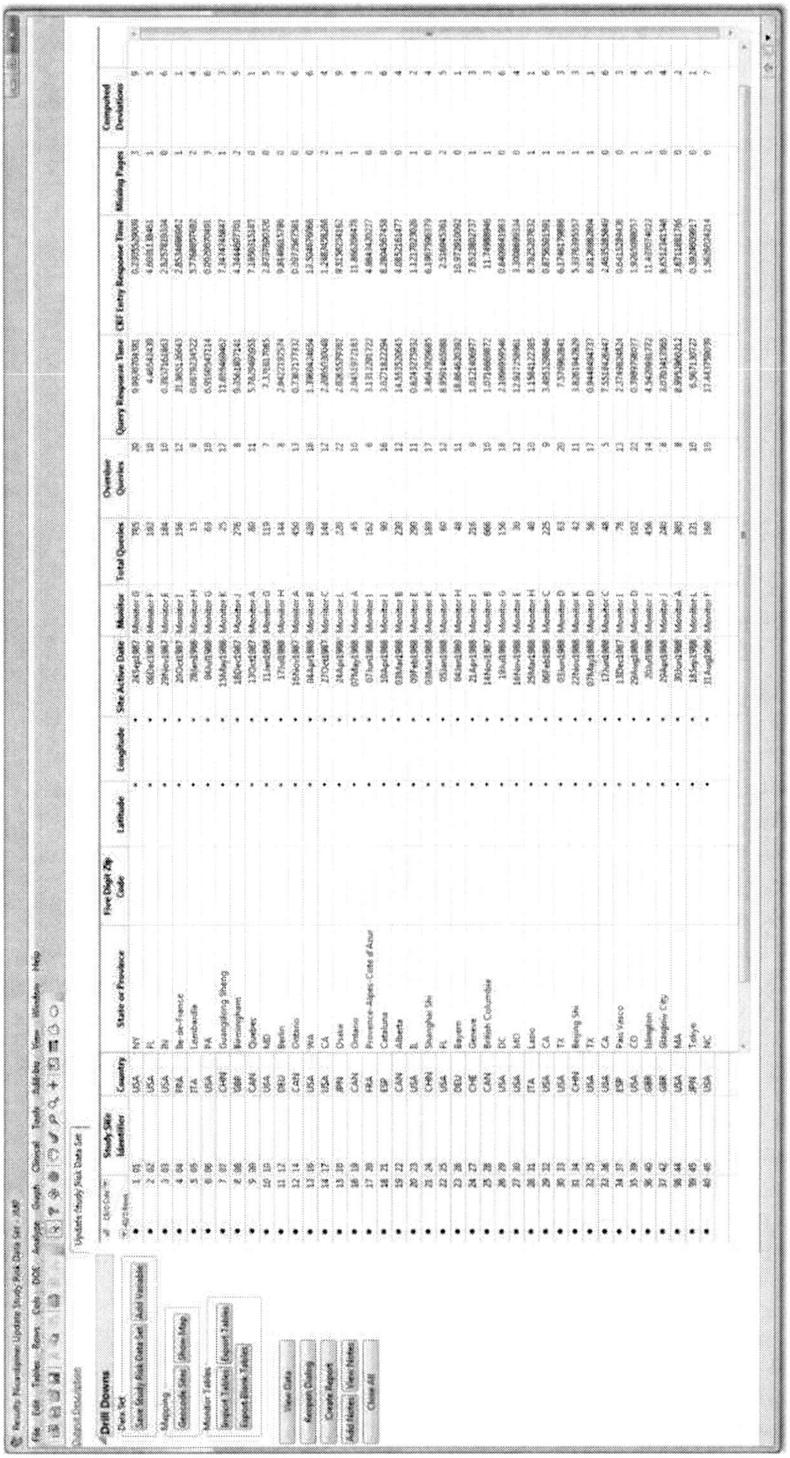

Figure 2.6 Typical Initial View of Study Risk Data Set

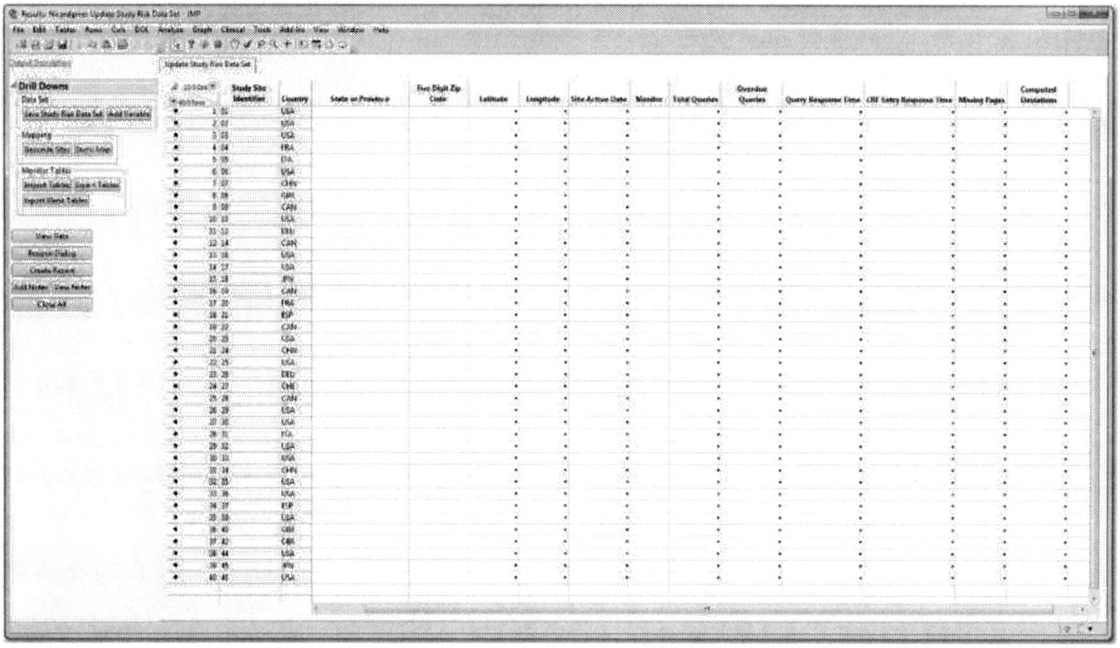

To make the data a bit more interesting, random site locations were generated across North America, Europe, and Asia (Figure 2.5 on page 33). Further, simulated data were developed to generate a **Site Active Date**, the primary **Monitor** responsible for the clinical site, and six common metrics used to assess site performance ordinarily not captured in the study database: **Total Queries**, **Overdue Queries**, **Query Response Time**, **CRF Entry Response Time**, **Missing Pages**, and **Computed Deviations**. The first 5 variables represent summary data that may be obtained from a DBMS; the last variable could summarize deviations derived from a statistical program or observed by a study monitor. Note that these data are captured at the site level, so that if the information is captured in a different manner, say at the subject level, the total should be used to represent the site. It is not required that any data be provided; the table may be left as is in Figure 2.6 on page 34.

Data entry can proceed in a number of ways, though how the user ultimately decides to complete the data table depends on how frequently the team wishes to update and review the RBM analysis, the resources available for data entry, and the team's ability to write scripts to retrieve data from their various data systems automatically (advanced scripting, however, is beyond the scope of this book). First, a single individual may have the responsibility for entering data from sources supplied by the study team; this could include listings, Excel spreadsheets, or SAS data sets (the latter two can be opened directly within JMP). The data can be typed in or copied and pasted into the data table. The **Study Risk Data Set** can be saved by clicking **Save Study Risk Data Set**.

Alternatively, a single individual can manage this data-updating process and allow others to share the data entry responsibility by using the **Import Tables**, **Export Tables**, and **Export Blank Tables** drill downs. To use these tools most effectively, it is ideal to have completed the **Monitor** column of the **Study Risk Data Set**. Once complete, select **Export Blank Tables**. The user will be asked to specify a directory in which to place a set of JMP data tables, one for each **Monitor**. To open a JMP table for viewing, say **Monitor A.jmp**, drag the table onto the JMP Clinical starter menu (Figure 2.7 on page 35). The contents of this file include the **Site Active Date**, the six predefined risk indicators, and any other risk indicators added using **Add Variable** for the set of sites for which Monitor A has primary responsibility. On a weekly basis, for example, the person responsible for updating the **Study Risk Data Set** can send out these blank JMP data tables to the monitor team. The monitor team can complete their data tables using their copies of JMP, and return them to the requesting individual. As a note, **Export Tables** generates a set of JMP data tables including the data that is currently available within the **Study Risk Data Set**. This can be a useful way to share current data with others.

Upon receipt, this individual can store the data tables in a single folder or directory. Selecting **Import Tables** and specifying the directory of completed tables will cause JMP Clinical to systematically cycle through each monitor data table to import and update the contents of the **Study Risk Data Set**. As an example, download **Updated JMP Tables.zip** from the companion website for this book, and place these JMP tables into a single folder. To witness the updating process, be sure that the **Study Risk Data Set** is scrolled to the right so that the 6 predefined risk indicators are onscreen. Use **Import Tables** to update the **Study Risk Data Set** (Figure 2.8 on page 36). The updated **Study Risk Data Set** is now ready for analysis using **Risk-Based Monitoring**. In order to return the **Study Risk Data Set** to its original form, download **Original JMP Tables.zip** and import these data tables (Figure 2.5 on page 33). The RBM analysis and review later in this chapter makes use of the original **Study Risk Data Set**. Failure to return the data table to its original form will result in minor differences in your output compared to the book.

Figure 2.7 Exported Blank Table for Monitor A

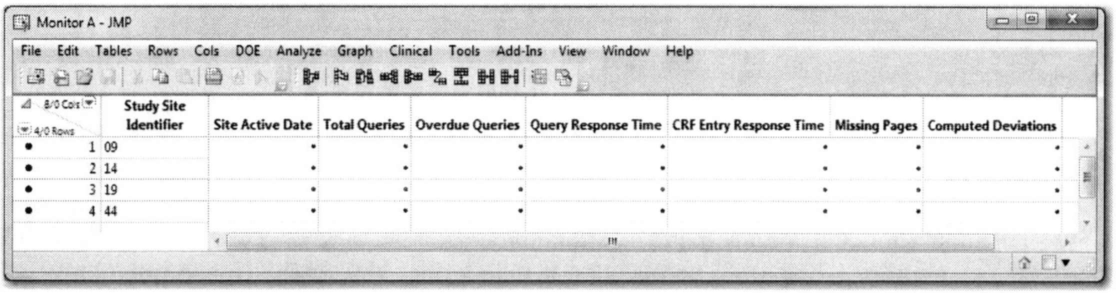

Figure 2.8 Post-Import Study Risk Data Table

	Study Site Identifier	Country	Total Queries	Overdue Queries	Query Response Time	CRF Entry Response Time	Missing Pages	Computed Deviations
1	01	USA	1020	111	20.993070438	3.2305529009	55	11
2	02	USA	322	68	9.46542439	6.6031138461	27	8
3	03	USA	309	55	4.3837161863	5.8257819334	24	7
4	04	FRA	285	55	34.985126643	3.8534686982	25	1
5	05	ITA	40	17	0.6879234522	5.7768857692	5	4
6	06	USA	93	22	7.9190547114	2.0020070491	13	6
7	07	CHN	51	26	11.855469462	7.3474745847	6	3
8	08	GBR	402	47	12.356180714	9.3444697701	21	7
9	09	CAN	272	87	16.782946096	10.195031519	45	3
10	10	USA	207	37	8.331817065	2.8737890326	14	5
11	12	DEU	219	43	6.9422192574	10.814661579	14	3
12	14	CAN	861	153	10.736717733	0.3972567581	68	8
13	16	USA	619	81	14.396042465	17.504676068	33	6
14	17	USA	245	43	4.2095030048	1.2483458288	25	4
15	18	JPN	331	70	7.0265579382	10.515823416	19	10
16	19	CAN	85	32	2.0451972183	12.866208478	14	4
17	20	FRA	236	42	5.1312291722	5.9843420227	21	3
18	21	ESP	137	43	4.0271822294	10.280456746	4	6
19	22	CAN	339	60	16.553520645	5.0852161477	30	4
20	23	USA	431	64	8.6243275932	1.1217023026	17	2
21	24	CHN	300	69	8.4642929685	8.1967596379	17	5
22	25	USA	118	43	11.959146509	3.516945361	23	5
23	26	DEU	89	18	20.864620392	10.972910092	6	1
24	27	CHE	352	63	7.0121406977	7.8523802737	28	4
25	28	CAN	1027	151	15.071666987	19.74988946	82	6
26	29	USA	213	47	3.1096959546	1.6409841983	10	6
27	30	USA	56	24	12.927258961	3.3008699334	3	4
28	31	ITA	62	20	2.1584122385	10.782520763	8	1
29	32	USA	478	101	16.495329885	0.8756501591	48	7
30	33	USA	111	37	8.570962841	7.1746179886	8	4
31	34	CHN	72	32	3.8261942829	5.3376395557	9	3
32	35	USA	104	36	2.9448494737	6.8126982804	10	1
33	36	USA	72	18	7.5518426447	4.4635285849	8	6
34	37	ESP	148	43	7.3749824824	0.6415289438	11	3
35	39	USA	193	52	4.7989798077	2.9265098057	24	4
36	40	GBR	639	91	11.542093177	16.437074022	53	6
37	42	GBR	344	46	14.070341356	10.651234155	14	5
38	44	USA	578	79	17.995296021	4.6711881786	33	4
39	45	JPN	333	54	10.567130727	5.3924609917	21	2
40	46	USA	290	61	24.443759039	4.5626024214	40	8

A few final notes about the importing and exporting process: First, any sites without a monitor assigned (i.e., **Monitor** is blank) will be exported to a JMP data table called **Unassigned.jmp**. This is true if only a handful or even all of the clinical sites are unassigned. Second, the importing process only updates values from the monitor tables that are non-missing (i.e., not a dot). In other words, if value(s) in the monitor tables are missing (say, a monitor had insufficient time to calculate the number of missing pages on a given week), the previous values in the **Study Risk Data Set** will be kept as is after import in order to maintain the most recent data available for analysis. My final comments have to do with whether your company is using JMP Clinical as a desktop product (local mode) or in server mode, connecting to a SAS metadata server to access study data. When operating locally, each analyst will have his or her own version of the **Study Risk Data Set** to develop and maintain (similar to how each person needs to register studies individually). The monitor tables can be placed in a network folder that can be accessed by everyone; each person will need to perform the importing process. When operating against a server, a single individual will update and maintain the **Study Risk Data Set**. Anyone who connects to the server with JMP Clinical and has the appropriate study permissions will have immediate access to the updated **Study Risk Data Set**. Analyses using **Risk-Based Monitoring** can proceed with the knowledge that the **Study Risk Data Set** is up to date.

Table 2.2 on page 37 summarizes **Monitor**, **Site Active Date**, and the six predefined risk indicators and their corresponding averages that can be used to supplement the study database for RBM analysis and review. Details for additional variables added to the **Study Risk Data Set** are described below.

Table 2.2   Risk Indicators from Supplemental Data

Grouping	Variable	Details
Site Identifiers	Monitor	Name of the individual with primary oversight of the clinical site
Enrollment Metrics	Site Active Date	Date the site becomes capable to enroll patients. This date is required to generate many of the enrollment indicators
Manually Entered Site Metrics	Total Queries	Number of queries at the clinical site
	Average Queries per Randomized Subject	Total Queries / Randomized
	Queries per PatientWeek	Total Queries / PatientWeeks
	Overdue Queries	Number of overdue queries at the clinical site, based on some predefined turnaround time for resolution
	Overdue Queries per Randomized Subject	Overdue Queries / Randomized
	Overdue Queries per PatientWeek	Overdue Queries / PatientWeeks
	Query Response Time	The average query response time. Units are unimportant, but should be consistent for all sites

Grouping	Variable	Details
	CRF Entry Response Time	The average CRF entry response time. Units are unimportant, but should be consistent for all sites.
	Missing Pages	Number of missing pages out of those expected
	Average Missing Pages per Randomized Subject	Missing Pages / Randomized
	Missing Pages per PatientWeek	Missing Pages / PatientWeeks
	Computed Deviations	Number of computed protocol deviations
	Average Computed Deviations per Randomized Subject	Computed Deviations / Randomized
	Computed Deviations per PatientWeek	Computed Deviations / PatientWeeks

Though JMP Clinical supplies six predefined risk indicators, it may be that other data are of interest and required for the RBM analysis. The user can add a new variable using the **Add Variable** drill down. Click this button, and add a new variable called **Computed Eligibility Violations**, for example (Figure 2.9 on page 38). This new variable will be added to the end of the data table.

Figure 2.9 Add Variable Drill Down

Let's simulate some sample data for eligibility violations. Right-click on **Computed Eligibility Violations** in the data table and go to **Formula > Random > Random Poisson**. Fill in the value 15 for **lambda** and click **OK** (Figure 2.10 on page 40). The data table is now populated with some random values which we will assume are the frequencies of eligibility violations at the site, though a more appropriate simulated value would consider the number of subjects randomized at the clinical site (Figure 2.11 on page 41, uses **Cols > Reorder Columns**). Note that the random values you obtain may differ from those in Figure 2.11, which will cause some differences in later analyses of this variable. If you would prefer to match analyses exactly, you can enter the data manually from Figure 2.11 or download **Computed Eligibility Violations.jmp** from the companion website. Drag the data table onto the JMP Clinical starter menu to open, then copy and paste the **Computed Eligibility Violations** column into the **Study Risk Data Set**. Click **Save Study Risk Data Set**. For advanced users, it is only possible to set a seed to obtain the same set of randomly-generated values when scripting with Random Reset().

You only need to add a particular variable once to make it available for future studies. In other words, if I were to register a new study to JMP Clinical to perform RBM analysis and review, **Computed Eligibility Violations** would be available in the **Study Risk Data Set** upon first running **Update Study Risk Data Set**. Similar to the other six predefined risk indicators, if this (or any other) added variable is not relevant to the current study, you may simply ignore it.

Figure 2.10 JMP Function Menu

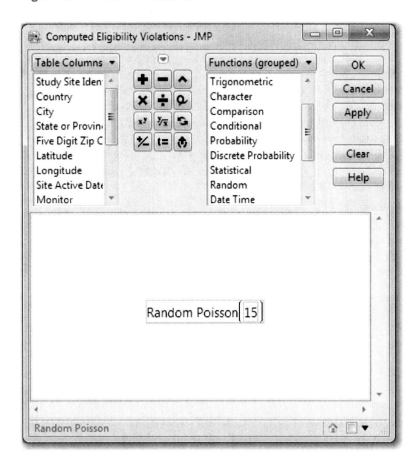

Figure 2.11 Random Data for Computed Eligibility Violations

	Study Site Identifier	Country	Computed Eligibility Violations
1	01	USA	13
2	02	USA	15
3	03	USA	15
4	04	FRA	11
5	05	ITA	14
6	06	USA	12
7	07	CHN	11
8	08	GBR	16
9	09	CAN	17
10	10	USA	12
11	12	DEU	14
12	14	CAN	11
13	16	USA	12
14	17	USA	12
15	18	JPN	16
16	19	CAN	16
17	20	FRA	16
18	21	ESP	13
19	22	CAN	14
20	23	USA	13
21	24	CHN	12
22	25	USA	17
23	26	DEU	13
24	27	CHE	13
25	28	CAN	14
26	29	USA	24
27	30	USA	19
28	31	ITA	13
29	32	USA	15
30	33	USA	14
31	34	CHN	14
32	35	USA	14
33	36	USA	19
34	37	ESP	24
35	39	USA	16
36	40	GBR	10
37	42	GBR	19
38	44	USA	14
39	45	JPN	23
40	46	USA	9

Table 2.3 on page 42 summarizes the variables that become available in the RBM analysis and review in **Risk-Based Monitoring** for each new variable added to the **Study Risk Data Set**, using **Computed Eligibility Violations** as an example. These new variables will be added to any newly generated risk threshold data sets when using **Define Risk Threshold Data Set**, but note that no risk thresholds or weight for overall indicators ("2.2.2 Overall Risk Indicators" on page 42) are initially defined. I'll show how to define risk thresholds for these user-added variables in "3.2 Defining Alternate Risk Thresholds and Actions" on page 80.

The next section discusses overall risk indicators and weighted combinations thereof.

Table 2.3   Risk Indicators from User-Added Variables

Grouping	Variable	Details
Manually Entered Site Metrics	Computed Eligibility Violations	Number of computed eligibility violations
	Computed Eligibility Violations per Randomized Subject	Computed Eligibility Violations / Randomized
	Computed Eligibility Violations per PatientWeek	Computed Eligibility Violations / PatientWeeks

2.2.2 Overall Risk Indicators

Up to this point, we have discussed the individual indicators of risk within our clinical trials. While it is important to understand how the sites perform for each of these particular indicators, it would be extremely useful to have some measure of safety and quality to assess the overall performance of the clinical sites. Such overall measures are straightforward to obtain by taking weighted averages of the individual risk indicators, centered and scaled so that no single variable will overwhelm the statistic. JMP Clinical provides five overall risk indicators to assess either the overall site performance by averaging all variables for which risk thresholds are defined (those columns with green, yellow, and red color), or the performance for subgroups of these risk indicators based upon the variable groupings: Enrollment Metrics, Disposition, Adverse Events, and Manually Entered. Overall indicators are provided if there is at least one variable with risk thresholds defined from the appropriate group with a non-zero weight assigned to determine how it will be combined with other risk indicators.

Table 2.4 on page 43 summarizes the overall variables in **Risk-Based Monitoring** based on the **Default Risk Threshold** data set. For individual risk indicators to contribute to overall indicators, the variables must be present in the **Site-** or **Country-Level Risk Indicators** data tables (Figure 2.3 on page 26), have risk thresholds and the **Weight for Overall Indicator** defined in the risk threshold data set applied to the RBM analysis, as well as have a non-zero standard deviation (i.e., the variable cannot be constant; otherwise it provides no information). I'll provide more formal definitions for risk thresholds and how they are combined in "3.2 Defining Alternate Risk Thresholds and Actions" on page 80.

Table 2.4 Overall Risk Indicators

Grouping	Variable	Details
Overall Risk Indicators	Overall Risk Indicator	Includes all variables below
	Overall Risk Indicator Adverse Events	Includes Deaths per PatientWeek, AEs per PatientWeek, SAEs per PatientWeek
	Overall Risk Indicator Disposition	Includes Percent Discontinued of Randomized Subjects, Average Inclusion or Exclusion Not Met per Randomized Subject, Average Deviations per Randomized Subject
	Overall Risk Indicator Enrollment	Includes Percent Screen Fail of Total Subjects, Missing Informed Consent, Randomized per Week Active
	Overall Risk Indicator Manually Entered	Includes Queries per PatientWeek, Overdue Queries per PatientWeek, Missing Pages per PatientWeek, Computed Deviations per PatientWeek, Query Response Time, CRF Entry Response Time

Users can add additional user-defined variables to the **Overall Risk Indicator Manually Entered** by supplying a non-zero **Weight for Overall Indicator**. Variables and their relative contributions can be modified for any overall indicator as described in Section 3.2.3.

2.2.3 Default Risk Thresholds

JMP Clinical ships with the **Default Risk Threshold** data set so that you can begin performing your RBM analyses as soon as possible, and to assist with developing your own risk threshold data sets. Table 2.5 on page 44 presents the variables for which thresholds are defined, describes the risk indicator category which determines the overall indicator subgroup to which a variable contributes, and notes which indicators do not contribute to the overall risk indicators. By default, all variables contribute to each overall risk indicator equally. Details on how to modify risk thresholds and values for the **Weight for Overall Indicator** are presented in Section 3.2.

Table 2.5 Default Risk Thresholds

Variable	Category	Green	Yellow	Red
Percent Screen Fail of Total Subjects	Enrollment	≤5% more or less than median value of all sites	>5% and ≤15% more or less than median value of all sites	>15% more or less than median value of all sites
Missing Informed Consent	Enrollment	No instances of a missing informed consent	-	At least one instance of a missing informed consent
Randomized per Week Active	Enrollment	≤5% more than median value of all sites	>5% and ≤15% more than median value of all sites	>15% more than median value of all sites
Observed Minus Expected Randomized*	Enrollment	<5 below the median of expected enrollment	≥5 and <10 below the median of expected enrollment	≥10 below the median of expected enrollment
Observed Minus Target Randomized*	Enrollment	<5 below the median of expected enrollment	≥5 and <10 below the median of expected enrollment	≥10 below the median of expected enrollment

Variable	Category	Green	Yellow	Red
Percent Discontinued of Randomized Subjects	Disposition	≤15% more or less than median value of all sites	>15% and ≤30% more or less than median value of all sites with an observed percentage of at least 3	>30% more or less than median value of all sites with an observed percentage of at least 3
Average Inclusion or Exclusion Not Met per Randomized Subject	Disposition	≤5% more or less than median value of all sites	>5% and ≤15% more or less than median value of all sites	>15% more or less than median value of all sites
Average Deviations per Randomized Subject	Disposition	≤5% more or less than median value of all sites	>5% and ≤15% more or less than median value of all sites	>15% more or less than median value of all sites
Percent Deaths of Randomized Subjects*	Adverse Events	≤5% more or less than median value of all sites	>5% and ≤10% more or less than median value of all sites	>10% more or less than median value of all sites
Deaths per PatientWeek	Adverse Events	≤5% more or less than median value of all sites	>5% and ≤10% more or less than median value of all sites	>10% more or less than median value of all sites
Average AEs per Randomized Subject*	Adverse Events	≤5% more or less than median value of all sites	>5% and ≤15% more or less than median value of all sites	>15% more or less than median value of all sites
AEs per PatientWeek	Adverse Events	≤5% more or less than median value of all sites	>5% and ≤15% more or less than median value of all sites	>15% more or less than median value of all sites
Average SAEs per Randomized Subject*	Adverse Events	≤5% more or less than median value of all sites	>5% and ≤15% more or less than median value of all sites	>15% more or less than median value of all sites

Variable	Category	Green	Yellow	Red
SAEs per PatientWeek	Adverse Events	≤5% more or less than median value of all sites	>5% and ≤15% more or less than median value of all sites	>15% more or less than median value of all sites
Average Queries per Randomized Subject*	Manually Entered	≤5% more than median value of all sites	>5% and ≤15% more than median value of all sites	>15% more than median value of all sites
Queries per PatientWeek	Manually Entered	≤5% more or less than median value of all sites	>5% and ≤15% more or less than median value of all sites with at least 2 queries per PatientWeek	>15% more or less than median value of all sites with at least 2 queries per PatientWeek
Average Overdue Queries per Randomized Subject*	Manually Entered	≤5% more than median value of all sites	>5% and ≤15% more than median value of all sites	>15% more than median value of all sites
Overdue Queries per PatientWeek	Manually Entered	≤5% more or less than median value of all sites	>5% and ≤15% more or less than median value of all sites with at least 2 overdue queries per PatientWeek	>15% more or less than median value of all sites with at least 2 overdue queries per PatientWeek
Average Missing Pages per Randomized Subject*	Manually Entered	≤5% more than median value of all sites	>5% and ≤15% more than median value of all sites	>15% more than median value of all sites
Missing Pages per PatientWeek	Manually Entered	≤5% more than median value of all sites	>5% and ≤15% more than median value of all sites	>15% more than median value of all sites
Average Computed Deviations per Randomized Subject*	Manually Entered	≤5% more than median value of all sites	>5% and ≤15% more than median value of all sites	>15% more than median value of all sites

Variable	Category	Green	Yellow	Red
Computed Deviations per PatientWeek	Manually Entered	≤5% more than median value of all sites	>5% and ≤15% more than median value of all sites	>15% more than median value of all sites
Query Response Time	Manually Entered	≤5% more than median value of all sites	>5% and ≤15% more than median value of all sites	>15% more than median value of all sites
CRF Entry Response Time	Manually Entered	≤5% more than median value of all sites	>5% and ≤15% more than median value of all sites	>15% more than median value of all sites
Overall Risk Indicator	Overall	≤15% more than median value of all sites	>15% and ≤30% more than median value of all sites	>30% more than median value of all sites
Overall Risk Indicator Adverse Events	Overall	≤15% more or less than median value of all sites	>15% and ≤30% more or less than median value of all sites	>30% more or less than median value of all sites
Overall Risk Indicator Disposition	Overall	≤15% more than median value of all sites	>15% and ≤30% more than median value of all sites	>30% more than median value of all sites
Overall Risk Indicator Enrollment	Overall	≤15% more than median value of all sites	>15% and ≤30% more than median value of all sites	>30% more than median value of all sites
Overall Risk Indicator Manually Entered	Overall	≤15% more than median value of all sites	>15% and ≤30% more than median value of all sites	>30% more than median value of all sites

Note: * Denotes variables that do not contribute to overall risk indicators.

2.2.4 Default Actions

While the criteria for risk thresholds to determine moderate or severe risk are important, the actions of the study team to either reduce or resolve risk are as important. These actions include, but are by no means limited to, calling the site, visiting the site to review procedures or source documents, reviewing additional data remotely or in greater detail, providing additional training, and assessing site resources and/or policies. The detail of such actions should be specified in the Integrated Quality Risk Management Plan (IQRMP) for moderate and severe levels of risk, as well as the expected outcome of any intervention. Section 8.1.4 of the Companion Guide to Risk Indicators of the TransCelerate Position Paper describes these actions in greater detail.

JMP Clinical specifies default actions for moderate (yellow) or severe (red) risk thresholds for a subset of the available risk indicators (Table 2.6 on page 48). Users are free to modify or add definitions to meet the criteria stated within the IQRMP and any modifications to these criteria stated within the study protocol or study-specific quality plan. Section 3.2.4 describes how to modify these criteria using **Define Risk Threshold Data Set**.

Table 2.6 Default Actions

Variable	Yellow	Red
Missing Informed Consent		Contact site principal investigator for missing informed consent.
Average Overdue Queries per Randomized Subject	Contact site coordinator for overdue queries.	Contact site principal investigator for overdue queries.
Overall Risk Indicator	Assess data remotely. Contact site coordinator.	Assess data remotely. Schedule on-site monitor visit.
Overall Risk Indicator Adverse Events	Assess data remotely. Contact site principal investigator.	Assess data remotely. Schedule on-site monitor visit.
Overall Risk Indicator Disposition	Assess data remotely. Contact site principal investigator.	Assess data remotely. Schedule on-site monitor visit.
Overall Risk Indicator Enrollment	Assess site resources.	Assess site resources. Contact site coordinator.
Overall Risk Indicator Manually Entered	Assess need for additional staff training. Contact site coordinator.	Schedule on-site monitor visit.

2.3 Geocoding Sites

While a data table of site-level risk indicators is useful for understanding individual site performance, such a presentation fails to consider the role geography and location may play in the collected data and their interpretation. For example, perhaps environmental differences at certain locations lend themselves to an increased incidence of certain safety issues. Sites within certain areas may have received training from a particular vendor, and trial monitors may be responsible for different sites based on their location. Further, countries may have differing standards and regulatory obligations that can impact the findings of an RBM analysis and review. These details are important for understanding why the risk may be elevated at certain sites, particularly if such patterns are seen across multiple clinical trials. Since DM.COUNTRY is a required variable for SDTM, it is rather straightforward to provide risk indicators at the country level, as well as geographic maps colored to highlight the risk of all sites within.

Pinpointing the exact site location, however, takes a bit of effort from the user. Data detailing the **City** of the clinical site, as well as the **State or Province** can be entered into **Update Study Risk Data Set** and, along with **Country**, can geocode the location of the site. Geocoding uses the provided geographic information to derive a latitude and longitude, which can be used to plot the site location on a map. The markers detailing the location of the clinical sites can be colored to display the severity of any of the available risk indicators. The aforementioned method works to geocode sites from any location worldwide. For example, in **Update Study Risk Data Set**, click **Geocode Sites**. After a brief pause while PROC GEOCODE performs its calculations, the **Latitude** and **Longitude** columns in the data table become populated with data (Figure 2.12 on page 50). The site locations are now available to view on a geographic map by selecting **Show Map** (Figure 2.13 on page 51). Of course, it is critically important to provide the appropriate spellings of geographic locations. **Map Geocoding Help** opens data tables for U.S. and international cities to assist users in their geocoding endeavors. Though CDISC recommends using International Organization for Standardization (ISO) 3166-1 alpha-3 country codes, ISO country names and alpha-2 codes should work to geocode sites.

Figure 2.12 Geographic Data in Study Risk Data Set

	Study Site Identifier	Country	City	State or Province	Five Digit Zip Code	Latitude	Longitude
1	01	USA	New York	NY		40.71417	-74.00639
2	02	USA	Orlando	FL		28.53806	-81.37944
3	03	USA	Indianapolis	IN		39.768329	-86.158052
4	04	FRA	Paris	Ile-de-France		48.856429	2.347557
5	05	ITA	Milano	Lombardia		45.4695	9.18172
6	06	USA	Philadelphia	PA		39.95222	-75.16417
7	07	CHN	Guangzhou	Guangdong Sheng		23.1189	113.261
8	08	GBR	Birmingham	Birmingham		52.497418	-1.907561
9	09	CAN	Montreal	Quebec		45.510616	-73.564501
10	10	USA	Baltimore	MD		39.29028	-76.6125
11	12	DEU	Berlin	Berlin		52.506394	13.412956
12	14	CAN	Toronto	Ontario		43.688621	-79.362983
13	16	USA	Seattle	WA		47.60639	-122.33083
14	17	USA	Los Angeles	CA		34.05222	-118.24278
15	18	JPN	Osaka	Osaka		34.666667	135.5
16	19	CAN	Ottawa	Ontario		45.395119	-75.713108
17	20	FRA	Marseille	Provence-Alpes-Cote d'Azur		43.288971	5.383338
18	21	ESP	Barcelona	Cataluna		41.404913	2.187463
19	22	CAN	Calgary	Alberta		51.016496	-114.066532
20	23	USA	Chicago	IL		41.85	-87.65
21	24	CHN	Shanghai	Shanghai Shi		31.2382	121.469
22	25	USA	Miami	FL		25.77389	-80.19389
23	26	DEU	Muenchen	Bayern		48.14208	11.578872
24	27	CHE	Geneve	Geneve		46.202551	6.141139
25	28	CAN	Vancouver	British Columbia		49.25266	-123.012386
26	29	USA	Washington	DC		38.895	-77.03667
27	30	USA	Saint Louis	MO		38.62722	-90.19778
28	31	ITA	Roma	Lazio		41.8753	12.5013
29	32	USA	San Diego	CA		32.71528	-117.15639
30	33	USA	Houston	TX		29.76306	-95.36306
31	34	CHN	Beijing	Beijing Shi		39.916691	116.396793
32	35	USA	Dallas	TX		32.78333	-96.8
33	36	USA	San Francisco	CA		37.775	-122.41833
34	37	ESP	Bilbao	Pais Vasco		43.259729	-2.933205
35	39	USA	Denver	CO		39.73917	-104.98417
36	40	GBR	London	Islington		51.525644	-0.094393
37	42	GBR	Glasgow	Glasgow City		55.85311	-4.218052
38	44	USA	Boston	MA		42.35833	-71.06028
39	45	JPN	Tokyo	Tokyo		35.698408	139.803956
40	46	USA	Charlotte	NC		35.22694	-80.84333

Figure 2.13 *Global Map of Clinical Sites*

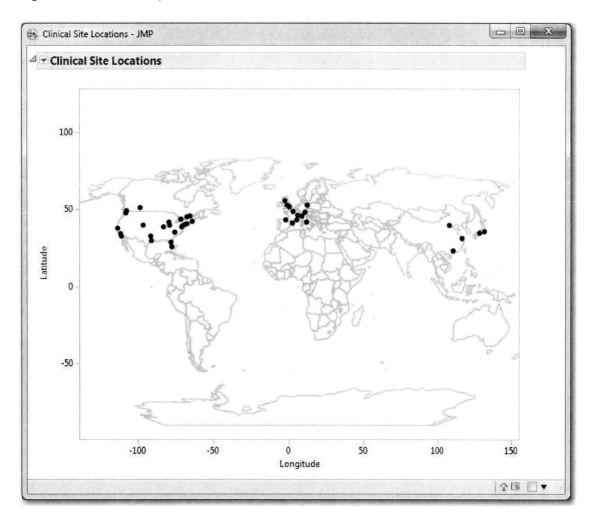

Providing **City** and **State or Province** is one way to geocode clinical sites. For U.S. sites, the **Five Digit Zip Code** can be used to save on some data entry efforts, in addition to providing a slightly more detailed geocoding for these centers. Providing zip codes can help distinguish map markers from sites within the same city that would ordinarily overlap. However, if postal zip codes are used to geocode at least one site, they must be used for all U.S. sites. Once **Five Digit Zip Code** is entered for all U.S. sites, the **Geocode Sites** drill down can be used to generate latitudes and longitudes. The final way of geocoding sites involves manually entering the latitude and longitude of the site into the data table. Though perhaps a bit tedious to enter, Global Positioning System (GPS) technologies are available in most handheld smartphones and from a variety of websites; the coordinates can be as specific as to geocode the waiting room within each center. While this method provides the most accurate locations of clinical sites, it is probably unnecessary in most applications. However, it could be very useful for trials with many clinical centers within the same city. In these situations, maps can be zoomed down to the city level with map backgrounds added (provided by OpenStreetMap). For example, right-click on the map shown in Figure 2.13 on page 51 and go to **Graph > Background Map > Street Map Service**. An example for sites in Florida is provided in Figure 2.14 on page 52.

Figure 2.14 Sites in Florida Using OpenStreetMap

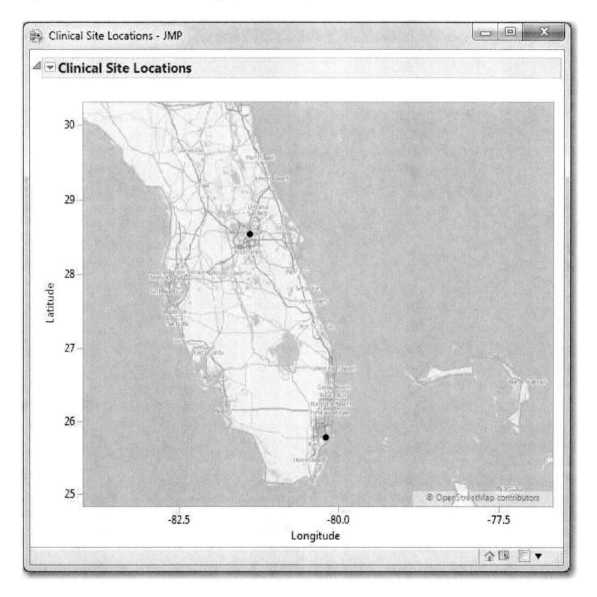

Table 2.7 on page 53 summarizes the site identifiers used to geocode clinical sites. Of course, similar to the other variables within the **Study Risk Data Set**, these data can be left blank. If so, the **Site-Level Risk Indicator Maps** tab from **Risk-Based Monitoring** will not be provided.

Table 2.7 Geographic Information

Grouping	Variable	Details
Site Identifiers	Country	DM.COUNTRY
	City	User provided city where site is located
	State or Province	User provided state, province, or district where site is located. U.S. states are supplied using the two-digit postal abbreviation.
	Five Digit Zip Code	For sites within the United States or from a U.S. territory, the user provided five-digit postal code.
	Latitude and Longitude	Latitude (North/South) or longitude (East/West). Derived from geographic information using **Geocode Sites** drill down or user provided.

2.4 Reviewing Risk Indicators

2.4.1 Site-Level Risk

2.4.1.1 Risk Indicator Data Table

Once supplemental data and geographic information have been supplied for the clinical sites through **Update Study Risk Data Set** (Section "2.2.1.2 Risk Indicators from Other Data Sources" on page 31 and Section "2.3 Geocoding Sites" on page 49) and a set of risk thresholds are defined through **Define Risk Threshold Data Set** (upcoming in Section 3.2), the user is now ready to perform an analysis of site performance using the **Risk-Based Monitoring** report. In actuality, however, the analyst is free to begin the analysis immediately, without adding any supplemental data, by taking advantage of the **Default Risk Threshold** data set. In this scenario, the RBM analysis and review will be limited solely to the data available within the study database, with no maps to display risks geographically based on site location. In this section, however, we'll review an RBM analysis of the **Nicardipine** study making use of the supplemental data available as part of the specially defined example (including the additional user-added variable **Computed Eligibility Violations**), and taking advantage of the latitude and longitude of sites based upon supplied geographic data to generate informative maps of risk. For now, we'll use the risk thresholds available in the **Default Risk Threshold** data set. I describe how to tailor risk thresholds to the needs of a specific clinical trial in Section 3.2.

Open the **Risk-Based Monitoring** dialog; be sure to select **Default Risk Threshold** and check the box to **Specify Target Enrollment** so that we can compare site performance in enrolling one patient for every two weeks (0.5 patients/week). Click **Run**. A tabbed report will summarize various aspects of the RBM analysis. Depending on whether sites were geocoded, the first two or three tabs summarize the RBM analysis at the site level (Figure 2.15 on page 55). Country-level analyses are presented on the remaining tabs; I describe this output in "2.4.2 Country-Level Risk" on page 66.

The **Site-Level Risk Indicators** tab displays a data table of the risk indicators supplied by the user through **Update Study Risk Data Set** or calculated directly from the study database. Listed above this data table, the user is reminded of the set of risk thresholds that were applied to the analysis. Each row indicates a unique clinical site (here, 40), and each column represents a different risk indicator. Row markers (the filled colored circles) to the left of the row numbers indicate the risk level for the variable selected in the **Risk Indicator** drill down on the left side of the dashboard, here the **Overall Risk Indicator**. Choosing other variables in this list will update the color of the row markers to correspond with the selected indicator. If risk thresholds were defined for a particular variable, the column will be colored in green, yellow, or red to indicate mild, moderate, or severe risk, respectively. The left panel adjacent to the data table indicates that there are 61 variables, categorized as Site Identifiers, Overall Indicators, Enrollment Metrics, Disposition, Adverse Events, and Manually Entered. The variable **Signal** (hidden in the data table as suggested by the eye goggles) is used to color the row markers, and the values of this variable change according to the variable selected in the **Risk Indicator** drill down. The other values in the data table correspond to frequencies, averages per randomized subject, or averages per **PatientWeek**; see Table 2.1 on page 28 and Table 2.4 on page 43 for descriptions of each variable. From here, the analyst has access to all of the safety and quality metrics at the site level, and can compare the performance of sites and evaluate risk severity to determine if intervention is necessary. Note that risk severity is not identified for any of the **Computed Eligibility Violation** variables. By default, thresholds are not automatically provided for new user-added variables. I'll define risk thresholds for the **Computed Eligibility Violation** variables in Section 3.2.2.

Figure 2.15 Subject-Level Risk Indicator Data Table

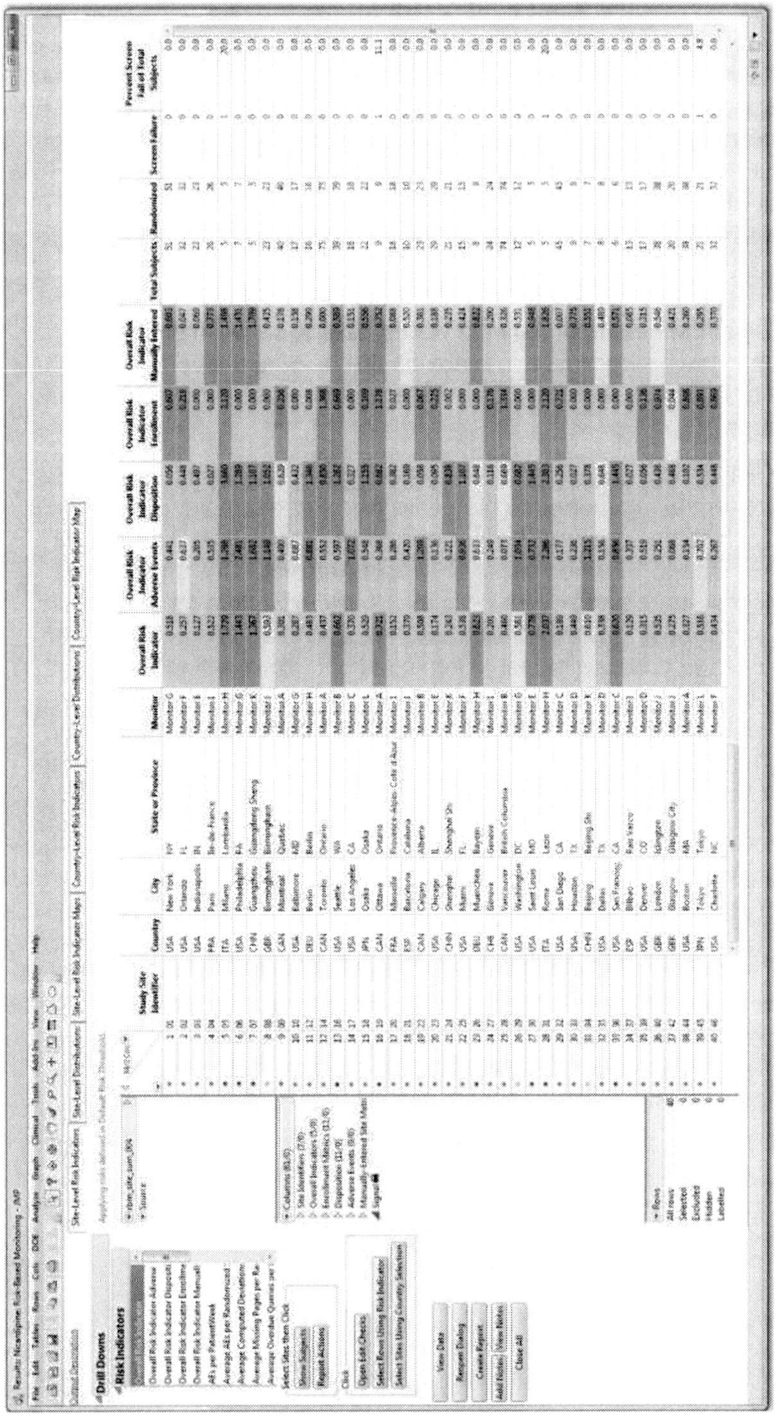

In Figure 2.15 on page 55, three and nine clinical sites were identified as having moderate and severe risk, respectively, for the **Overall Risk Indicator** based on the risk thresholds defined at the tail end of Table 2.5 on page 44. From the **Site-Level Distributions** tab (discussed below), we can easily determine the median value of this variable as 0.47148 from the summary table of the appropriate histogram. If interested, users can add an additional variable to the data table to calculate the exact percentage above or below this median value. For example, right-click on a blank column, select **New Column**, and name it **Percentage Above the Overall Median**. Right-click again and choose **Formula**, and use the formula editor to define this value as ((:Overall Risk Indicator - Col Quantile(:Overall Risk Indicator, 0.5)) / Col Quantile(:Overall Risk Indicator, 0.5)) * 100. In addition, you can select sites based on the risk severity for further analysis by using the **Select Rows Using Risk Indicator**. This drill down allows you to select any combination of green, yellow, or red severities based on the variable selected in the **Risk Indicator** drill down. Selecting rows using this drill down, the cursor, or **Rows > Rows Selection** will highlight or emphasize these particular sites in the histograms and maps on the other site-level tabs. I'll show some additional functionality of selection in Sections "2.4.2 Country-Level Risk" on page 66 and "2.4.3 Subject-Level Risk" on page 70

The **Report Actions** drill down will open a summary table containing the suggested actions by clinical site for any risk indicators displaying moderate or severe risk. For example, using **Select Rows Using Risk Indicator** to select the sites with severe risk for the **Overall Risk Indicator** provides the table in Figure 2.16 on page 57 when clicking **Report Actions**. This table shows the recommended actions for all moderate and severe risks for all risk indicators for these sites. Using **Analyze > Distribution** and choosing **Study Site Identifier**, **Label** and **Recommended Action** for **Y, Columns** can be used to collapse the table into a set of distinct actions (Figure 2.17 on page 58). Selecting a particular site, such as site 36, will highlight the **Recommended Action** histogram to show how many times an action was suggested. Alternatively, users may add a **Local Data Filter** through the red triangle menu and **Script > Local Data Filter** to subset histograms to particular sites or actions. An example is described in "2.4.1.2 Distributions" on page 60.

Figure 2.16 Site Actions for Moderate or Severe Risk

Study Site Identifier	Variable	Value	Label	Recommended Action
1 05	AVOVERQ	1.6	Average Overdue Queries per Randomized Subject	Contact site principal investigator for overdue queries.
2 05	OVERALL	1.729	Overall Risk Indicator	Assess data remotely. Schedule onsite monitor visit.
3 05	OVERALLA	1.288	Overall Risk Indicator Adverse Events	Assess data remotely. Schedule onsite monitor visit.
4 05	OVERALLD	3.66	Overall Risk Indicator Disposition	Assess data remotely. Schedule onsite monitor visit.
5 05	OVERALLE	2.12	Overall Risk Indicator Enrollment	Assess site resources. Contact site coordinator.
6 05	OVERALLM	1.498	Overall Risk Indicator Manually Entered	Schedule onsite monitor visit.
7 06	AVOVERQ	1.429	Average Overdue Queries per Randomized Subject	Contact site principal investigator for overdue queries.
8 06	OVERALL	1.443	Overall Risk Indicator	Assess data remotely. Schedule onsite monitor visit.
9 06	OVERALLA	2.481	Overall Risk Indicator Adverse Events	Assess data remotely. Schedule onsite monitor visit.
10 06	OVERALLD	1.289	Overall Risk Indicator Disposition	Assess data remotely. Schedule onsite monitor visit.
11 06	OVERALLM	1.431	Overall Risk Indicator Manually Entered	Schedule onsite monitor visit.
12 07	AVOVERQ	3.4	Average Overdue Queries per Randomized Subject	Contact site principal investigator for overdue queries.
13 07	OVERALL	1.367	Overall Risk Indicator	Assess data remotely. Schedule onsite monitor visit.
14 07	OVERALLA	1.682	Overall Risk Indicator Adverse Events	Assess data remotely. Schedule onsite monitor visit.
15 07	OVERALLD	1.107	Overall Risk Indicator Disposition	Assess data remotely. Schedule onsite monitor visit.
16 07	OVERALLM	1.709	Overall Risk Indicator Manually Entered	Schedule onsite monitor visit.
17 16	OVERALL	0.662	Overall Risk Indicator	Assess data remotely. Schedule onsite monitor visit.
18 16	OVERALLD	1.282	Overall Risk Indicator Disposition	Assess data remotely. Schedule onsite monitor visit.
19 16	OVERALLE	0.669	Overall Risk Indicator Enrollment	Assess site resources. Contact site coordinator.
20 16	OVERALLM	0.589	Overall Risk Indicator Manually Entered	Schedule onsite monitor visit.
21 19	AVOVERQ	1.111	Average Overdue Queries per Randomized Subject	Contact site principal investigator for overdue queries.
22 19	OVERALL	0.721	Overall Risk Indicator	Assess data remotely. Schedule onsite monitor visit.
23 19	OVERALLD	0.682	Overall Risk Indicator Disposition	Assess data remotely. Schedule onsite monitor visit.
24 19	OVERALLE	1.178	Overall Risk Indicator Enrollment	Assess site resources. Contact site coordinator.
25 19	OVERALLM	0.752	Overall Risk Indicator Manually Entered	Schedule onsite monitor visit.
26 26	AVOVERQ	1.375	Average Overdue Queries per Randomized Subject	Contact site principal investigator for overdue queries.
27 26	OVERALL	0.623	Overall Risk Indicator	Assess data remotely. Schedule onsite monitor visit.
28 26	OVERALLA	0.633	Overall Risk Indicator Adverse Events	Assess data remotely. Contact site principal investigato
29 26	OVERALLD	0.648	Overall Risk Indicator Disposition	Assess data remotely. Contact site principal investigato
30 26	OVERALLM	0.822	Overall Risk Indicator Manually Entered	Schedule onsite monitor visit.
31 30	AVOVERQ	2.4	Average Overdue Queries per Randomized Subject	Contact site principal investigator for overdue queries.
32 30	OVERALL	0.778	Overall Risk Indicator	Assess data remotely. Schedule onsite monitor visit.
33 30	OVERALLA	0.732	Overall Risk Indicator Adverse Events	Assess data remotely. Schedule onsite monitor visit.
34 30	OVERALLD	1.445	Overall Risk Indicator Disposition	Assess data remotely. Schedule onsite monitor visit.
35 30	OVERALLM	0.948	Overall Risk Indicator Manually Entered	Schedule onsite monitor visit.
36 31	AVOVERQ	2	Average Overdue Queries per Randomized Subject	Contact site principal investigator for overdue queries.
37 31	OVERALL	2.037	Overall Risk Indicator	Assess data remotely. Schedule onsite monitor visit.
38 31	OVERALLA	2.286	Overall Risk Indicator Adverse Events	Assess data remotely. Schedule onsite monitor visit.
39 31	OVERALLD	2.383	Overall Risk Indicator Disposition	Assess data remotely. Schedule onsite monitor visit.
40 31	OVERALLE	2.12	Overall Risk Indicator Enrollment	Assess site resources. Contact site coordinator.
41 31	OVERALLM	1.826	Overall Risk Indicator Manually Entered	Schedule onsite monitor visit.
42 36	AVOVERQ	0.833	Average Overdue Queries per Randomized Subject	Contact site principal investigator for overdue queries.
43 36	OVERALL	0.62	Overall Risk Indicator	Assess data remotely. Schedule onsite monitor visit.
44 36	OVERALLA	0.856	Overall Risk Indicator Adverse Events	Assess data remotely. Schedule onsite monitor visit.
45 36	OVERALLD	1.445	Overall Risk Indicator Disposition	Assess data remotely. Schedule onsite monitor visit.
46 36	OVERALLM	0.571	Overall Risk Indicator Manually Entered	Schedule onsite monitor visit.

When **Site Active Date** is supplied using **Update Study Risk Data Set**, the study team is able to calculate and compare enrollment at the clinical sites in two ways (Figure 2.18 on page 59). **Expected Randomized** is calculated from **Weeks Active** using a rate based on combining the enrollment performance of all sites together. **Target Randomized** is calculated based on the supplied target enrollment rate from the study dialog (default half-a-subject per week). Using the thresholds supplied in **Default Risk Threshold**, severity is determined for **Observed Minus Expected Randomized** or **Observed Minus Target Randomized** by comparing how far the difference in subjects is below the overall median for the sites. Examining over-performance for these variables is worthwhile to identify sites that could potentially enroll patients not meeting eligibility criteria.

Figure 2.17 Distributions of Site Actions

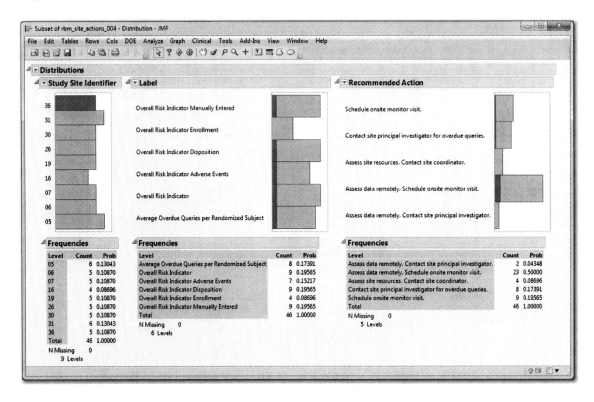

In "2.2.1.1 Risk Indicators from CDISC-Formatted Data Sets" on page 22, I indicated that it is possible to perform sensitivity analyses for data from the AE, DV, or IE data domains using the **WHERE** statements in the **Risk-Based Monitoring** dialog. To illustrate the use of these statements, let's perform an analysis where we limit the AEs to those that are of moderate or severe severity. For the **Filter to Choose Adverse Events for RBM** in the dialog, write the text AESEV in ('MODERATE','SEVERE'). This will limit the analysis to moderate and severe AEs and provide us an opportunity to examine whether there is any under- or over-reporting of more severe events. Compare the results of Figure 2.19 on page 60 with those of Figure 2.4 on page 27. The large number of cells indicating severe risk could be interpreted as sites having large variability around the median of the AE risk indicators. Of course, any number of sensitivity analyses could be performed for adverse events, protocol deviations, or inclusion or exclusion criteria not met. This would require some understanding of the contents of the CDISC data tables; statisticians, programmers, and/or data managers from the study team can assist with variable names and values.

Figure 2.18   Enrollment Risk Indicators

	Study Site Identifier	Country	Site Active Date	Weeks Active	Randomized per Week Active	Expected Randomized	Observed Minus Expected Randomized	Target Randomized	Observed Minus Target Randomized
1	01	USA	24Sep1987	102.14	0.499	31	20	51	0
2	02	USA	06Dec1987	91.71	0.349	28	4	46	-14
3	03	USA	29Nov1987	92.71	0.248	28	-5	46	-23
4	04	FRA	20Oct1987	98.43	0.264	30	-4	49	-23
5	05	ITA	28Jan1988	84.14	0.059	26	-21	42	-37
6	06	USA	04Jul1988	61.57	0.114	19	-12	31	-24
7	07	CHN	15May1988	68.71	0.073	21	-16	34	-29
8	08	GBR	18Dec1987	90.00	0.256	28	-5	45	-22
9	09	CAN	13Oct1987	99.43	0.402	30	10	50	-10
10	10	USA	11Jan1988	86.57	0.196	27	-10	43	-26
11	12	DEU	17Jul1988	59.71	0.268	18	-2	30	-14
12	14	CAN	16Nov1987	94.57	0.793	29	46	47	28
13	16	USA	04Apr1988	74.57	0.523	23	16	37	2
14	17	USA	27Oct1987	97.43	0.185	30	-12	49	-31
15	18	JPN	24Apr1988	71.71	0.307	22	0	36	-14
16	19	CAN	07May1988	69.86	0.129	21	-12	35	-26
17	20	FRA	07Jun1988	65.43	0.275	20	-2	33	-15
18	21	ESP	10Apr1988	73.71	0.136	23	-13	37	-27
19	22	CAN	03Mar1988	79.14	0.291	24	-1	40	-17
20	23	USA	09Feb1988	82.43	0.352	25	4	41	-12
21	24	CHN	03Mar1988	79.14	0.265	24	-3	40	-19
22	25	USA	05Jan1988	87.43	0.172	27	-12	44	-29
23	26	DEU	04Jan1989	35.29	0.227	11	-3	18	-10
24	27	CHE	21Apr1988	72.14	0.333	22	2	36	-12
25	28	CAN	14Nov1987	94.86	0.780	29	45	47	27
26	29	USA	19Jul1988	59.43	0.202	18	-6	30	-18
27	30	USA	16Nov1988	42.29	0.118	13	-8	21	-16
28	31	ITA	25Mar1988	76.00	0.066	23	-18	38	-33
29	32	USA	06Feb1988	82.86	0.543	25	20	41	4
30	33	USA	03Jun1988	66.00	0.136	20	-11	33	-24
31	34	CHN	22Nov1988	41.43	0.169	13	-6	21	-14
32	35	USA	07May1988	69.86	0.115	21	-13	35	-27
33	36	USA	17Jun1988	64.00	0.094	20	-14	32	-26
34	37	ESP	13Dec1987	90.71	0.143	28	-15	45	-32
35	39	USA	29Aug1988	53.57	0.317	16	1	27	-10
36	40	GBR	20Jul1988	59.29	0.641	18	20	30	8
37	42	GBR	29Apr1988	71.00	0.282	22	-2	36	-16
38	44	USA	30Jun1988	62.14	0.611	19	19	31	7
39	45	JPN	18Sep1988	50.71	0.414	16	5	25	-4
40	46	USA	31Aug1988	53.29	0.601	16	16	27	5

One last important feature to highlight for the **Site-Level Risk Indicator** tab involves the ability for users to make and save notes at the site level using the **Add Notes** and **View Notes** drill downs. These particular drill downs are described in detail in Section 6.4 and can be used to create notes at the analysis, patient, or record level. In **Risk-Based Monitoring**, you can create date-time-stamped notes for clinical sites to remind yourself of any important details that may be pertinent to future analyses or for mention in the clinical study report. To add a note for one or more sites, select the rows of interest and click **Add Notes**. Once you click **OK**, the note is saved to a database to be retrieved at a later time. Any available notes can be accessed by selecting the relevant sites and clicking **View Notes**. For example, use **Select Rows Using Risk Indicator** to select sites where the **Overall Risk Indicator** is severe (red). Click **Add Notes** and enter some text (Figure 2.20 on page 61). Click **OK** to save the note. With the same set of sites selected, clicking **View Notes** will open a data table of all the available notes for these sites. Though a single note was entered, a separate note record was entered for each of the selected clinical sites.

Figure 2.19 Risk Indicators for Moderate and Severe Adverse Events

	Study Site Identifier	Country	Total AEs on Study	Average AEs per Randomized Subject	AEs per PatientWeek	Total SAEs on Study	Average SAEs per Randomized Subject	SAEs per PatientWeek
1	01	USA	277	5.43	3.14	51	1.00	0.58
2	02	USA	44	1.38	0.75	13	0.41	0.22
3	03	USA	72	3.13	1.94	10	0.43	0.27
4	04	FRA	106	4.08	2.23	25	0.96	0.53
5	05	ITA	29	5.80	4.41	3	0.60	0.46
6	06	USA	59	8.43	5.43	15	2.14	1.38
7	07	CHN	26	5.20	4.04	4	0.80	0.62
8	08	GBR	157	6.83	3.83	36	1.57	0.88
9	09	CAN	104	2.60	1.75	22	0.55	0.37
10	10	USA	58	3.41	2.03	20	1.18	0.70
11	12	DEU	71	4.44	2.66	17	1.06	0.64
12	14	CAN	160	2.13	1.23	29	0.39	0.22
13	16	USA	61	1.56	0.96	5	0.13	0.08
14	17	USA	103	5.72	3.29	25	1.39	0.80
15	18	JPN	41	1.86	1.07	8	0.36	0.21
16	19	CAN	16	1.78	1.14	5	0.56	0.36
17	20	FRA	32	1.78	1.06	12	0.67	0.40
18	21	ESP	58	5.80	3.01	13	1.30	0.67
19	22	CAN	132	5.74	3.23	55	2.39	1.35
20	23	USA	114	3.93	2.51	17	0.59	0.37
21	24	CHN	63	3.00	1.91	10	0.48	0.30
22	25	USA	94	6.27	3.68	19	1.27	0.74
23	26	DEU	15	1.88	1.06	2	0.25	0.14
24	27	CHE	80	3.33	1.83	20	0.83	0.46
25	28	CAN	329	4.45	2.37	68	0.92	0.49
26	29	USA	40	3.33	1.81	17	1.42	0.77
27	30	USA	3	0.60	0.36	1	0.20	0.12
28	31	ITA	15	3.00	3.50	5	1.00	1.17
29	32	USA	215	4.78	2.77	42	0.93	0.54
30	33	USA	27	3.00	2.01	6	0.67	0.45
31	34	CHN	41	5.86	3.54	16	2.29	1.38
32	35	USA	41	5.13	2.87	5	0.63	0.35
33	36	USA	25	4.17	2.16	1	0.17	0.09
34	37	ESP	37	2.85	1.61	3	0.23	0.13
35	39	USA	60	3.53	2.59	16	0.94	0.69
36	40	GBR	55	1.45	0.83	16	0.42	0.24
37	42	GBR	54	2.70	1.62	13	0.65	0.39
38	44	USA	150	3.95	2.33	20	0.53	0.31
39	45	JPN	24	1.14	0.72	1	0.05	0.03
40	46	USA	138	4.31	2.46	17	0.53	0.30

2.4.1.2 Distributions

The **Site-Level Distributions** tab displays histograms and box plots for a majority of the available risk indicators (Figure 2.21 on page 62). Here the analyst can determine the distributions of the risk indicators, identify any outliers (colored according to the **Risk Indicator** drill down), and review quantiles and summary statistics. For example, while nine centers are considered to have a risk level of severe (red) for the **Overall Risk Indicator**, only four of these sites appear to be outliers according to the corresponding box plot. These sites can be identified by hovering over the markers with the cursor, or by dragging the cursor and selecting the set of sites that highlights the corresponding rows of the data table. The red triangles next to each variable name allow the analyst to tailor display options, generate additional plots, perform statistical tests, generate confidence intervals, or fit parametric or non-parametric curves to the data. Further, the histograms themselves can be modified in useful ways. For example, right-click on the y-axis of **Overall Risk Indicator**, and go to **Axis Settings > Scale > Log** to present the values with a log-transformation (Figure 2.22 on page 62). (Transformed variables can be added to the data table for analysis by going to **Cols > Transform**.) Other histogram options are available when right-clicking inside the border of each figure.

Figure 2.20 Site-Level Notes

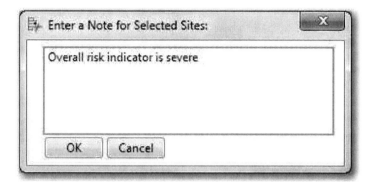

Similar to our data table example above, all histograms are linked; selecting a particular bar or outlier(s) will highlight the corresponding observations in the other histograms. For example, selecting the cells for Monitor A and B (while holding the Shift key) displays where these sites fall relative to the other sites in all histograms (Figure 2.23 on page 63). Histogram selection also highlights the relevant rows of the data table, as well as the site markers on any maps.

62 Chapter 2 / Risk-Based Monitoring: Basic Concepts

Figure 2.21 Distributions of Overall Risk Indicators

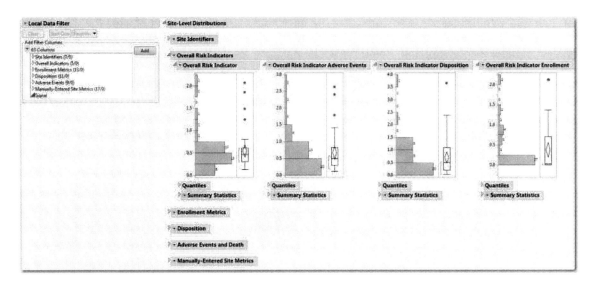

Figure 2.22 Log-Transform of Overall Risk Indicator

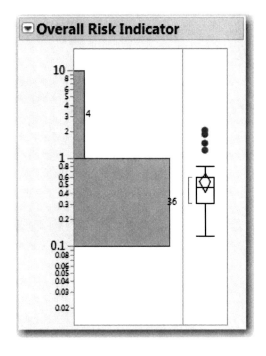

Figure 2.23 Highlight Sites for Monitor A and B

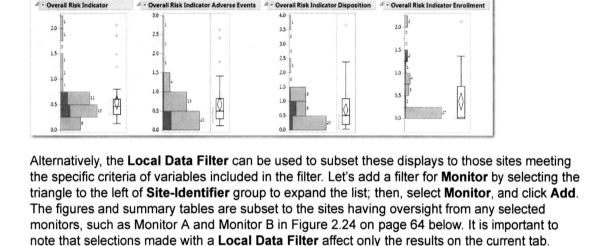

Alternatively, the **Local Data Filter** can be used to subset these displays to those sites meeting the specific criteria of variables included in the filter. Let's add a filter for **Monitor** by selecting the triangle to the left of **Site-Identifier** group to expand the list; then, select **Monitor**, and click **Add**. The figures and summary tables are subset to the sites having oversight from any selected monitors, such as Monitor A and Monitor B in Figure 2.24 on page 64 below. It is important to note that selections made with a **Local Data Filter** affect only the results on the current tab.

Figure 2.24 Subset Histograms to Sites of Monitor A and B Using Local Data Filter

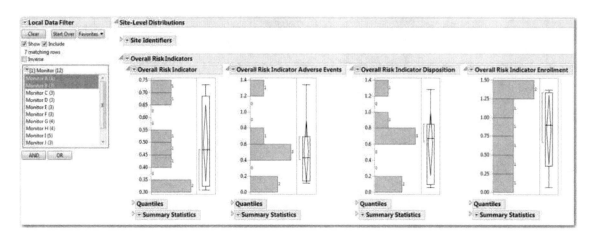

Interested (and thorough) users may be curious as to why some variables were excluded from the histograms. This was done intentionally to prevent the tab from feeling too overcrowded by summarizing every risk indicator within the data table. However, any variable in the data table can be summarized by going to **Analyze > Distribution** and selecting as many variables for **Y, Columns** as are of interest. Such an approach may also be useful to summarize certain variables side-by-side for ease of review. A **Local Data Filter** can be added to these new results by clicking the red triangle menu beside **Distributions**, and clicking **Script > Local Data Filter**.

2.4.1.3 Maps

The **Site-Level Risk Indicator Maps** tab is available whenever at least one site has latitude and longitude available. A global map presents the risk for geocoded sites for the variable highlighted in the **Risk Indicator** drill down (Figure 2.25 on page 65). Users can cycle through the **Risk Indicator** list to easily view the risk severity for all variables geographically. Below, the severity for the **Overall Risk Indicator** is displayed. No obvious pattern is discernable for the United States, though eastern sites in Europe tend to be of high overall risk. Site performance in China appears to be discordant, while Japan appears to be performing well overall. A **Local Data Filter** on the tab makes it possible to subset to observations meeting important criteria, such as those sites belonging to a particular set of monitors. Further, maps of the following countries are available if at least one clinical site is present within the country: Canada, China, France, Germany, Great Britain, Italy, Japan, Spain, and the United States (Figure 2.26 on page 66). Country maps can be useful to distinguish between sites with close proximity to one another.

Figure 2.25 Site Locations Colored by Severity of Overall Risk Indicator

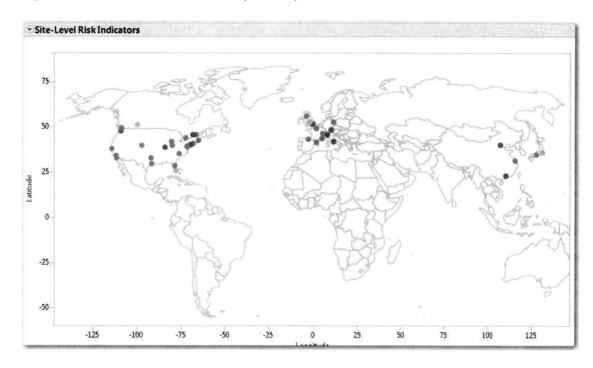

66 Chapter 2 / Risk-Based Monitoring: Basic Concepts

Figure 2.26 Site Locations in the United States Colored by Severity of Overall Risk Indicator

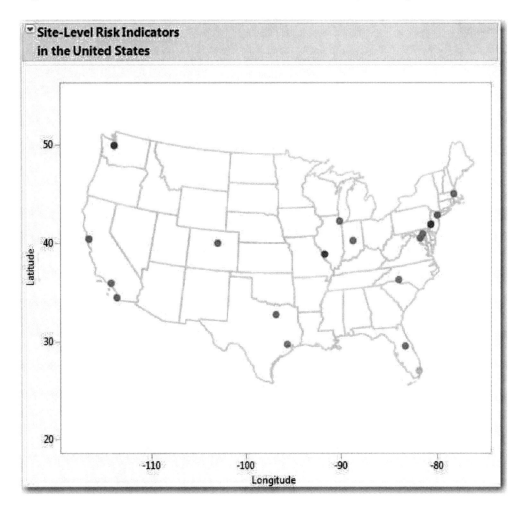

2.4.2 Country-Level Risk

Most of the previous discussion for site-level risk indicators is applicable here for risk summarized at the country level. Why bother summarizing risk at the country level at all? For one reason, each country has its own set of rules, regulations, and standards to follow that can impact the safety of trial subjects or the quality of data collected from clinical sites. Even countries that are members of the European Union and regulated by the European Medicines Agency have subtle nuances in their regulatory practices. For this reason, it is a good idea to get some assessment of the performance of sites for each country. Furthermore, this country-level assessment may suggest that all sites within a particular country would benefit from an intervention, not just the one or two sites bordering on severe risk. Summarizing data at the country level also helps in the interpretation of sites with a small number of randomized subjects. The findings from these sites may be extremely variable, with summary values appearing extreme due to the small number of patients under study. Summarizing at the country level may help the sponsor ultimately decide to intervene with a site with a small number of randomized patients.

Figure 2.27 Country-Level Risk Indicator Data Table

	Country	Overall Risk Indicator	Overall Risk Indicator Adverse Events	Overall Risk Indicator Disposition	Overall Risk Indicator Enrollment	Overall Risk Indicator Manually Entered	Total Sites	Total Subjects	Randomized	Screen Failure	Percent Screen Fail of Total Subjects
1	CAN	0.283	0.388	0.267	0.897	0.029	5	221	221	1	0.5
2	CHE	0.202	0.482	0.023	0.185	0.098	1	24	24	0	0.0
3	CHN	0.307	0.415	0.181	0.000	0.376	3	33	33	0	0.0
4	DEU	0.360	0.213	0.724	0.000	0.493	2	24	24	0	0.0
5	ESP	0.171	0.193	0.032	0.000	0.240	2	23	23	0	0.0
6	FRA	0.317	0.430	0.045	0.000	0.412	2	44	44	0	0.0
7	GBR	0.315	0.112	0.060	0.322	0.457	3	81	81	0	0.0
8	ITA	1.766	2.028	2.915	1.595	1.501	2	10	10	2	20.0
9	JPN	0.628	1.421	0.711	0.443	0.280	2	43	43	1	2.3
10	USA	0.073	0.100	0.023	0.068	0.069	18	403	403	0	0.0

The **Country-Level Risk Indicators** data table (Figure 2.27 on page 67) presents a similar summary as for the site-level table, though here each row represents a unique country where one or more sites is active (here, 10). In general, values for variables are summed across all sites within a country to get appropriate totals. Appropriate averages per randomized subject or patient weeks are then calculated. The two exceptions to this rule are **Query Response Time** and **CRF Entry Response Time**; the mean value from all sites within a country is used as the country-level estimate. Risk severity is determined by applying the same rules as were used for the site-level table from the **Risk Threshold Data Set** selected in the dialog. Of course, the rules are applied somewhat differently since it is the means, medians, and standard deviations of values at the site level that are used to determine individual indicator severity, as well as the severity for any overall indicators. The **Country-Level Distributions** tab provides a similar set of histograms in order to review the distribution, quantiles, and summary statistics of the risk indicators. When **Add Notes** and **View Notes** drill downs are applied to observations from this data table or values selected from distributions or maps, a country-level note is saved to the note database.

The most notable differences between the site- and country-level summaries involve the geographic maps (Figure 2.28 on page 68). Here, only a global map is provided and entire countries are colored based on the risk severity of the variable selected in the **Risk Indicators** drill down. Based on the map of the **Overall Risk Indicator** shown below, we may conclude that Italy and Japan have severe overall risk, while Germany shows moderate risk. However, reviewing the country-level analysis without considering the site-level analysis can be extremely misleading. If we compare the results of Figure 2.15 on page 55 and Figure 2.25 on page 65 to Figure 2.27 on page 67 and Figure 2.28 on page 68, we appear to have a very unusual result. The Japanese sites show mild risk overall when considered individually, but severe risk when combined! What is going on? The **Overall Risk Indicator** for Japan is more or less in the same ballpark compared to the values of the individual Japanese sites. If you examine the other sites that have severe overall risk, you'll notice that a large number of these centers reside in the United States. However, when the U.S. sites are collapsed into a single unit, the extremeness of these sites is hidden by the sites with mild risk. The Japanese risk that didn't appear extreme when analyzing the sites separately now appears to be extreme. Given that these sites have 20 randomized subjects apiece and mild risk at the site level, I would interpret this country-level risk to be an anomaly. In general, I would err on the side of site results unless the sites themselves have particularly low enrollment. In any event, any discordances between site- and country-level results should be understood. The severity for Italy makes sense given the performance of the individual sites. Once you are comfortable with Section 3.2 as an exercise, reanalyze the **Nicardipine** data by centering all risk indicators currently using the median around the mean instead, then, see if this unusual result for Japan is repeated.

Figure 2.28 Countries Colored by Severity of Overall Risk Indicator

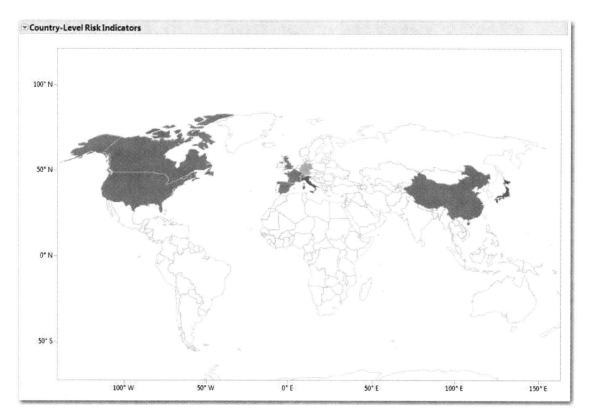

Because the country-level maps and distributions are linked to the **Country-Level Risk Indicators** data table, selecting rows from the data table will highlight observations in the figures and vice versa. Risk severity selections made using the **Select Rows Using Risk Indicators** drill down also selects the rows from the country-level data table that meet the chosen criteria. An additional drill down, **Select Sites Using Country Selection**, is a way for the user to easily identify sites from any country selections made in the country-level results tabs. For example, suppose I highlight the outlier (Italy) from the **Overall Risk Indicator** histogram on the **Country-Level Distributions** tab. If I click **Select Sites Using Country Selection**, the Italian sites are selected in the site-level table. From here, I can click **Show Subjects** to review the patients from these sites. As another example, click **Select Rows Using Risk Indicators** to select sites and countries with severe risk for the **Overall Risk Indicator**. This highlights all rows from both tables with a red marker. Now if I click **Select Sites Using Country Selection**, the site selection is updated to select all sites from countries with severe risk for the **Overall Risk Indicator**. For **Nicardipine**, this will select all sites from Italy and Japan.

2.4 Reviewing Risk Indicators 69

Figure 2.29 Data Table of Subjects from Sites with Severe Overall Risk

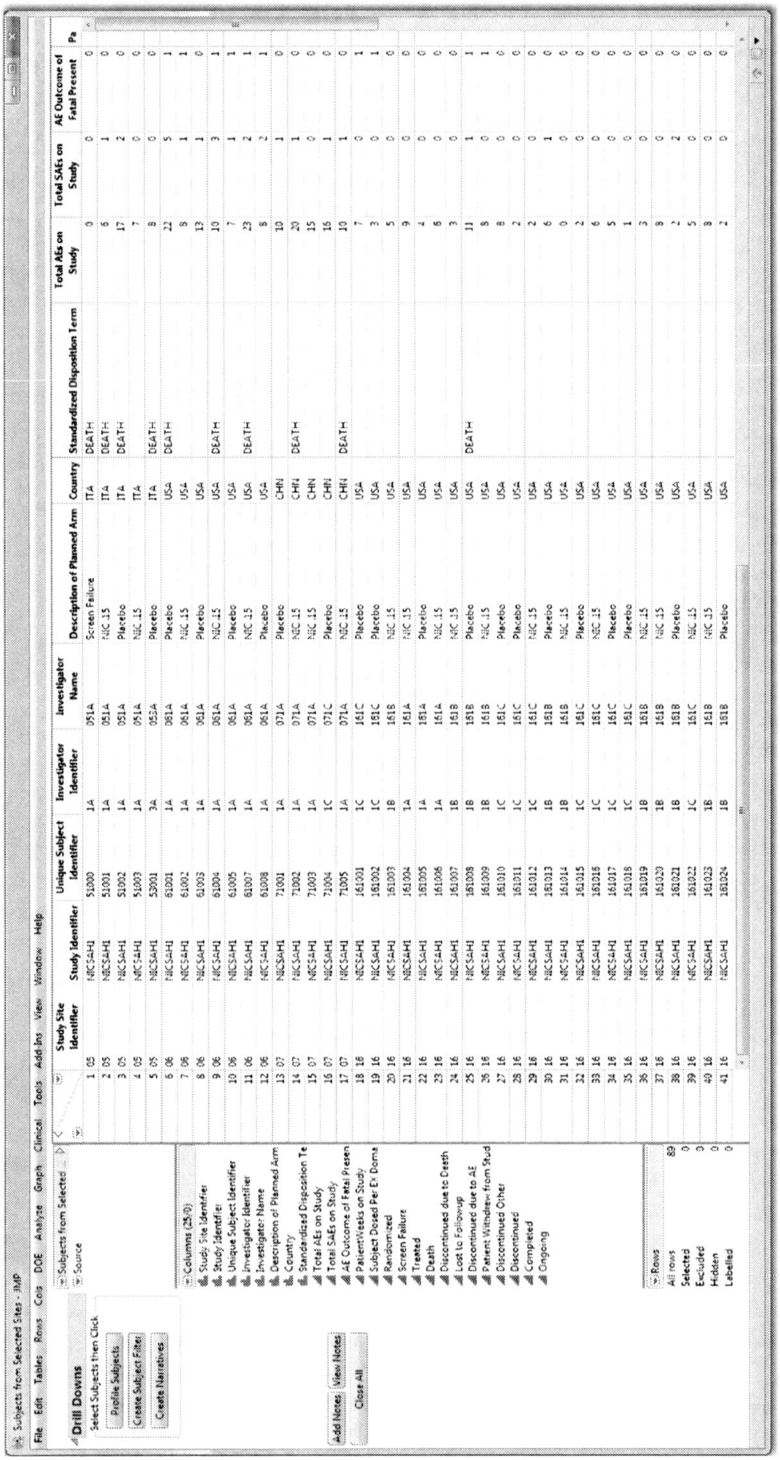

2.4.3 Subject-Level Risk

A benefit of performing your RBM analysis and review within JMP Clinical is the readily available access to subject-level data through the study database. RBM drill downs allow the analyst to easily select "interesting" sites, or sites from "interesting" countries; these selections can be used to identify and display subjects from these study centers by clicking the **Show Subjects** drill down. For example, use **Select Rows Using Risk Indicator** to select rows with severe risk for the **Overall Risk Indicator**. Now click **Show Subjects** to open a data table of the 89 study participants from these nine clinical sites (Figure 2.29 on page 69). The data table contains risk indicators derived from the study database at the subject level. This table can be reviewed to identify the specific subjects meeting criteria reported at the site level (for example, which subjects died or discontinued the study). A data filter (**Rows > Data Filter**) can be used to help highlight observations of interest. For selected rows, the analyst can explore the subjects in further detail by generating graphical and tabular patient profiles using **Patient Profiles** (Figure 2.30 on page 70) to summarize the totality of a subject's data in a single tab, or by generating written summaries of their adverse events using **AE Narrative**.

Figure 2.30 Patient Profile

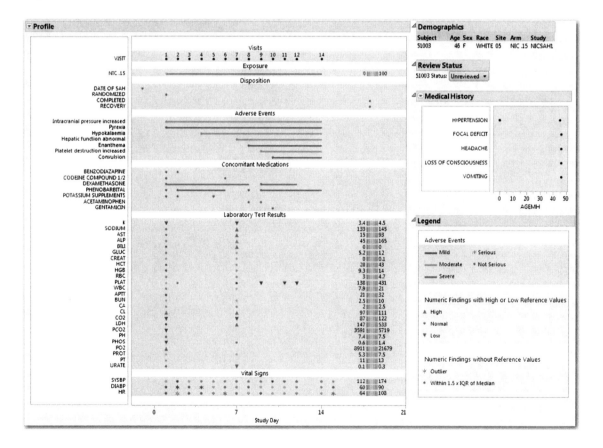

Additionally, a subject filter can be generated using **Create Subject Filter** to subset analyses performed in other JMP Clinical reports to the currently selected set of subjects by choosing a **Subject Filter** on the **Filter** tab of report dialogs. For example, use **Ctrl-A** to select all 89 patients. Click **Create Subject Filter** and name the filter **Subjects from Severe Sites** (Figure 2.31 on page 71). Go to the Clinical Starter and to **Events > Adverse Events > AE Distribution**, and select this newly created subject filter on the **Filters** tab (Figure 2.32 on page 71). Also select the **Count multiple occurrences of an event per subject** option on the **General** tab to display all adverse events for these patients. **Run** the analysis to get the output shown in (Figure 2.33 on page 72). This graphical summary allows you to review the individual adverse events from the individuals from sites identified with severe overall risk.

Figure 2.31 Subject Filter for Patients from High-Risk Sites

Figure 2.32 Subject Filter on AE Distribution Filter Tab

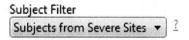

72 Chapter 2 / Risk-Based Monitoring: Basic Concepts

Figure 2.33 Summary of Adverse Events for Patients from High-Risk Sites

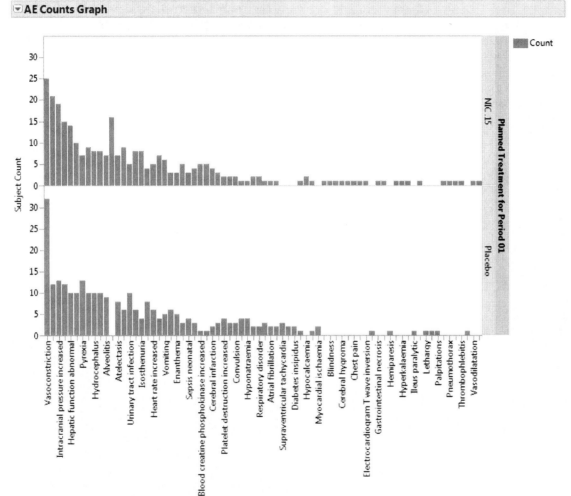

Finally, the drill down **Open Edit Checks** opens a PDF document that summarizes any inconsistencies in the subject-level risk indicators defined from the study database. For the **Nicardipine** trial, the following issues were reported:

1. There are several instances where a subject had a fatal AE but was not marked as having discontinued the trial due to death. A summary listing is provided for any variables that indicate whether a subject has died. Given that there are several locations within the CDISC model where death can be reported, these data should be examined for consistency and for proper counts of subjects that died. For the RBM analysis, a subject is considered to have died if he or she is reported to have died in any CRF source.

2. There are four instances where subjects were simultaneously identified as being randomized and screen failures. For this trial, this is due to a data quality issue described in Section 1.5.

3. These same four individuals were listed as randomized, but were never treated with the study drug. While this may happen from time to time, it is important to identify these individuals in the clinical study report.

Other edit checks will look for the following occurrences:

1. Subjects treated with study therapy in the EX domain that are not listed as randomized.

2. Subjects who are considered to have simultaneously completed and discontinued the trial.

3. Subjects missing the date that informed consent was signed.

Additional analyses of subject-level data will be presented in Chapters 4 and 5. In order to open a data table of all available subjects, use **Ctrl-A** to easily select all sites and push **Show Subjects**.

2.5 Final Thoughts

This chapter served as a thorough overview of the basic RBM capabilities available in JMP Clinical. From my perspective, RBM is a topic with which the entire clinical development team should be familiar. Everyone is (and should be) responsible for the quality of the data, particularly those data important to the well-being of the study participants and the integrity and validity of the study conclusions. By proactively identifying important data, implementing regular review of these data, and defining appropriate risk management, important safety and quality issues can be identified and corrected while the trial is ongoing. This ongoing review should minimize any last-minute surprises for the study team, including discovering any unfortunate findings after locking the study database. Furthermore, by developing detailed protocols and thorough standard operating procedures and by providing sufficient training with a particular focus to these data, safety and quality issues can be prevented or minimized from the start. RBM can result in not only more efficient trial oversight, but also substantial savings to the clinical trial sponsor. Ideally, these savings will benefit consumers and be used to develop innovative therapies in areas of unmet medical need. In the next chapter, I hope to inspire the imagination of analysts and reviewers to the wide variety of additional functionality available.

References

1. TransCelerate BioPharma Inc. (2013). Position paper: Risk-based monitoring methodology. Available at: http://transceleratebiopharmainc.com/.

2   Venet D, Doffagne E, Burzykowski T, Beckers F, Tellier Y, Genevois-Marlin E, Becker U, Bee V, Wilson V, Legrand C & Buyse M. (2012). A statistical approach to central monitoring of data quality in clinical trials. *Clinical Trials* 9: 705–713.

3   Eisenstein EL, Lemons PW, Tardiff BE, Schulman KA, Jolly MK & Califf RM. (2005). Reducing the costs of phase III cardiovascular clinical trials. *American Heart Journal* 149: 482–488.

4   Tantsyura V, Grimes I, Mitchel J, Fendt K, Sirichenko S, Waters J, Crowe J & Tardiff B. (2010). Risk-based source data verification approaches: pros and cons. *Drug Information Journal* 44: 745–756.

5   International Conference of Harmonisation. (1996). E6: Guideline for Good Clinical Practice. Available at: http://www.ich.org/fileadmin/Public_Web_Site/ICH_Products/Guidelines/Efficacy/E6_R1/Step4/E6_R1__Guideline.pdf

6   Funning S, Grahnén A, Eriksson K & Kettis-Linblad A. (2009). Quality assurance within the scope of good clinical practice (GCP): what is the cost of GCP-related activities? A survey with the Swedish association of the pharmaceutical industry (LIF)'s members. *The Quality Assurance Journal* 12: 3–7.

7   Grieve AP. (2012). Source data verification by statistical sampling: issues in implementation. *Drug Information Journal* 46: 368–377.

8   US Food & Drug Administration. (2013). Guidance for industry: Oversight of clinical investigations - a risk-based approach to monitoring. Available at: http://www.fda.gov/downloads/Drugs/.../Guidances/UCM269919.pdf.

9   Baigent C, Harrell FE, Buyse M, Emberson JR & Altman DG. (2008). Ensuring trial validity by data quality assurance and diversification of monitoring methods. *Clinical Trials* 5: 49–55.

10 Bakobaki JM, Rauchenberger M, Joffe N, McCormack S, Stenning S & Meredith S. (2012). The potential for central monitoring techniques to replace on-site monitoring: findings from an international multi centre clinical trial. *Clinical Trials* 9: 257–264.

11 Nielsen E, Hyder D & Deng C. (2014). A data-driven approach to risk-based source data verification. *Therapeutic Innovation and Regulatory Science* 48: 173–180.

12 European Medicines Agency. (2011). Reflection paper on risk based quality management in clinical trials. Available at: http://www.ema.europa.eu/docs/en_GB/document_library/Scientific_guideline/2013/11/WC500155491.pdf.

13 Medicines and Healthcare Products Regulatory Agency. (2011). Risk-adapted approaches to the management of clinical trials of investigational medicinal products. Available at: http://www.mhra.gov.uk/home/groups/l-ctu/documents/websiteresources/con111784.pdf.

14 US Food and Drug Administration. (2012). PDUFA reauthorization performance goals and procedures fiscal years 2013 through 2017. http://www.fda.gov/downloads/ForIndustry/UserFees/PrescriptionDrugUserFee/UCM270412.pdf.

15 US Food and Drug Administration. (2013). Position Statement for Study Data Standards for Regulatory Submissions. http://www.fda.gov/ForIndustry/DataStandards/StudyDataStandards/ucm368613.htm.

16 Zink RC & Mann G. (2012). On the importance of a single data standard. *Drug Information Journal* 46: 362–367.

Appendix

Walk-through of the RBM Reports

With the numerous reports and analyses available for RBM, as well as the interconnectivity between them, navigating the JMP Clinical Starter can initially seem overwhelming. There are also slight differences in how the software performs depending on whether you are working locally or connecting to a SAS metadata server to access study data. The Clinical Starter menu can be simplified to include only those analyses that are needed using **General Utilities > Customize Clinical Starter**. Below is a brief walk-through to help get you started, particularly if you are new to JMP Clinical.

1. First, register a study using **Add Study from Folders**. This will make your study available in the **Current Study** drop-down on the JMP Clinical starter menu, as well as the **Study** drop down within report dialogs. If working with JMP Clinical in server mode, someone else likely has the responsibility for registering and updating studies with JMP Clinical. The study will be available to you in drop-down menus when you start JMP Clinical if you have the appropriate study permissions. The individual managing study data may make use of **Add Study from CDI** or **Add Study from Metadata Libraries**. Studies are typically registered using the most recent data snapshot.

2. Use **Update Study Risk Data Set** to add data that is external to the database to the **Study Risk Data Set**. These data will be incorporated into the RBM analysis. Further, additional geographic information for the clinical sites can be added here in order to geocode the site locations. This allows you to view risk indicators geographically in **Risk-Based Monitoring**. If working in server mode, the **Study Risk Data Set** is shared among all individuals with access to this particular study, so only one individual or set of individuals should be responsible for maintaining it. This step can be skipped if there is no additional supplemental data or interest in site locations.

3. If there is interest in defining a novel set of risk thresholds or recommended actions to apply to the RBM analysis, this can be performed in **Define Risk Threshold Data Set**. If working with JMP Clinical in server mode, any risk threshold data sets created will be accessible to everyone with access to the server. This step can be skipped initially or in its entirety; the RBM analyses can make use of the **Default Risk Threshold** data set. Note that any added

variables using **Add Variable** in **Update Study Risk Data Set** will neither have risk thresholds defined nor be incorporated into any overall risk indicators until defined by the user.

4. Perform your analysis and review using **Risk-Based Monitoring**, selecting the appropriate **Risk Threshold Data Set**. If working in server mode, any notes made using **Add Notes** will be available to anyone with access to the study. The RBM analysis can be performed multiple times, applying different risk threshold data sets.

5. If a new data snapshot of the study database becomes available, go to **Studies > Manage Studies > Update Study Data and Metadata** and provide the new **SDTM** and/or **ADaM Folders**. If working in server mode, this step will likely be handled for you.

6. If a new supplemental data becomes available, add or update this data using **Update Study Risk Data Set**. If working in server mode, this step may be managed by another individual or set of individuals.

7. Each time data is refreshed, perform your analysis and review using **Risk-Based Monitoring**, selecting the appropriate risk threshold data sets.

Definitions of Risk Indicators and Important Terms in Pseudo-code

JMP Clinical has very minimal requirements to perform analyses related to RBM. Variables will be written below as domain.domain-variable to be explicit. For example, USUBJID from the DM domain will be written as DM.USUBJID. When a term can apply to multiple domains, "xx" will be used to imply any two-letter domain code. To show flexibility to new users of CDISC standards, JMP Clinical accepts values that do not strictly adhere to CDISC or its accepted controlled terminology. Such deviations may not necessarily be acceptable to regulatory agencies requiring data sets submitted using CDISC standards.

JMP Clinical requires the following data and domains:

1. Demography (DM): Study Site Identifier (DM.SITEID) and Country (DM.COUNTRY);

2. Disposition (DS): Standardized Disposition Term (DS.DSDECOD);

3. Either of the following to calculate the number of weeks a patient has been on-study (PatientWeeks):

    a. Subject Visits (SV): Start Date/Time of Visit (SV.SVSTDTC) and if available End Date/Time of Visit (SV.SVENDTC), or

    b. One or more Findings domains, such as Vital Signs (VS), EG (ECG test results), or LB (laboratory test results): Date/Time of Measurements (xx.xxDTC).

Though not required, the Exposure (EX) domain can be used to determine if a subject was treated with study therapy.

Data from the following domains are used if available to generate additional risk indicators:

1. Adverse Events (AE)
2. Protocol Deviations (DV)
3. Inclusion/Exclusion Criterion Not Met (IE).

Below are important details for how JMP Clinical defines various terms necessary to conduct the analysis.

1. Subjects are considered RANDOMIZED if there is at least one record from DS where index(upcase(DS.DSDECOD), "RANDOMIZED") is true.

2. Depending on the available information, subjects are considered SCREEN FAILURES if

    a. upcase(DM.ARM) or upcase(DM.ACTARM) = "SCREEN FAILURE", or

    b. upcase(DS.EPOCH) = "SCREENING" and upcase(DS.DSCAT) = "DISPOSITION EVENT" and upcase(DS.DSDECOD) ^= "COMPLETED", or

    c. upcase(DS.EPOCH) = "SCREENING" and upcase(DS.DSDECOD) ^= "COMPLETED", or

    d. subject is not randomized based on (1).

3. To determine if subjects COMPLETED the trial, a SAS WHERE statement can be supplied in the dialog to pick the appropriate DS records (this statement should also pick the records that indicate whether a subject alternatively is marked as DISCONTINUED or WITHDRAWN). If this where statement is supplied and upcase(DS.DSDECOD) = "COMPLETED" then the subject will be considered to have completed the trial. Otherwise, based on the available variables the subject is considered to have completed the trial if

    a. upcase(DS.EPOCH) = "TREATMENT" and upcase(DS.DSCAT) = "DISPOSITION EVENT" and upcase(DS.DSDECOD) = "COMPLETED", or

    b. upcase(DS.EPOCH) = "TREATMENT" and upcase(DS.DSDECOD) = "COMPLETED", or

    c. upcase(DS.DSCAT) = "DISPOSITION EVENT" and upcase(DS.DSDECOD) = "COMPLETED".

4. Similarly, subjects will be considered to have DISCONTINUED or WITHDRAWN if this where statement is supplied and upcase(DS.DSDECOD) ^= "COMPLETED". Otherwise, based on the available variables the subject is considered to have discontinued the trial if

    a. upcase(DS.EPOCH) = "TREATMENT" and upcase(DS.DSCAT) = "DISPOSITION EVENT" and upcase(DS.DSDECOD) ^= "COMPLETED", or

    b. upcase(DS.EPOCH) = "TREATMENT" and upcase(DS.DSDECOD) ^= "COMPLETED", or

    c. upcase(DS.DSCAT) = "DISPOSITION EVENT" and upcase(DS.DSDECOD) ^= "COMPLETED"

5. Randomized subjects who have neither completed nor discontinued the trial are considered ONGOING.

6. Depending on the available information, subjects are considered TREATED if

   a. Date/Time of First Study Treatment (DM.RFXSTDTC) is non-missing, or

   b. If records are available for the subject in the EX domain, or

   c. upcase(DM.ACTARM) not in ("SCREEN FAILURE","NOT TREATED","NOT ASSIGNED"), or

   d. upcase(DM.ARM) not in ("SCREEN FAILURE","NOT TREATED","NOT ASSIGNED"), or

   e. subject is randomized based on (1).

7. An adverse event will be considered serious (an SAE) if upcase(AE.AESER) in ("Y","YES").

8. An AE will be considered fatal if upcase(AE.AEOUT) in ("FATAL","DEATH") or upcase(AESDTH) in ("Y","YES").

9. A subject will be considered to have signed informed consent if Date/Time of Informed Consent (DM.RFICDTC) is non-missing.

10. A subject will be considered to have died if

    a. Date/Time of Death (DM.DTHDTC) is non-missing, or

    b. upcase(DM.DTHFL) in ("Y", "YES"), or

    c. the subject experienced a fatal AE (as defined in 8), or

    d. the subject discontinued the trial with upcase(DS.DSDECOD) in ("DEATH","DIED","DEAD").

11. Subject discontinuations or withdrawals are separated into various reasons for discontinuation:

    a. If upcase(DS.DSDECOD) in ("DEATH","DIED","DEAD"), then the subject is considered to have Discontinued Due to Death, or

    b. If upcase(DS.DSDECOD) in ("LOST TO FOLLOW-UP","LOST TO FOLLOWUP","LOST TO FOLLOW UP","LTFU"), then the subject is considered to be Lost to Followup, or

    c. If upcase(DS.DSDECOD) in ("ADVERSE EVENT","AE"), then the subject is considered to have Discontinued Due to Adverse Event, or

    d. If upcase(DS.DSDECOD) in ("WITHDRAWAL BY SUBJECT","SUBJECT WITHDRAWAL","WITHDREW CONSENT","SUBJECT WITHDREW CONSENT"), then the Patient Withdrew from Study, or

    e. The subject will be considered to have Discontinued Other.

3

Risk-Based Monitoring: Customizing the Review Experience

3.1 Introduction 79
3.2 Defining Alternate Risk Thresholds and Actions 80
    3.2.1 Thresholds for Individual and Overall Risk Indicators 80
    3.2.2 Thresholds for User-Added Risk Indicators 87
    3.2.3 Weights for Overall Risk Indicators 89
    3.2.4 Defining Actions for Elevated Risk 93
3.3 Performing Additional Statistical and Graphical Analyses 95
    3.3.1 Statistical Analyses 95
    3.3.2 Graphing 103
3.4 Creating JMP Scripts and Add-Ins 108
    3.4.1 New Discontinuation Variables and Risk Thresholds 108
    3.4.2 New Adverse Event Variables and Figures 114
    3.4.3 Analyses at the Monitor Level 120
3.5 Final Thoughts 124
References 124

3.1 Introduction

In Chapter 2, I introduced the basic RBM analysis and review functionality of JMP Clinical, making use of the **Default Risk Threshold** data set. Further, I described how JMP Clinical makes use of available data in the study database in order to define risk indicators for RBM, and summarized how users can supplement this data with other information available from DBMS, programs, or spreadsheets. In this chapter, I describe some more advanced topics. In "3.2 Defining Alternate Risk Thresholds and Actions" on page 80, we spend some time understanding how the default risk thresholds described in Table 2.5 are defined for use by the analysis programs within the **Default Risk Threshold** data set. Next, we examine how the overall risk indicators are combined from the individual risk indicators. In either case, the examples provided should give you the ability to modify and define new risk thresholds to tailor the analysis in a way most appropriate for your clinical trials. In Sections "3.3 Performing Additional Statistical and Graphical Analyses" on page 95 and "3.4 Creating JMP Scripts and Add-Ins" on page 108, I illustrate how the analyst can easily customize their review experience to generate additional statistical and graphical analyses, and by using JMP scripts and Add-Ins to define new risk indicators from the RBM site- or country-level data tables, respectively.

3.2 Defining Alternate Risk Thresholds and Actions

3.2.1 Thresholds for Individual and Overall Risk Indicators

JMP Clinical is shipped with a predefined set of risk thresholds contained within the **Default Risk Threshold** data set. This set of thresholds can be applied or selected by choosing **Default Risk Threshold** in either the **Risk-Based Monitoring** or **Define Risk Threshold Data Set** reports. Default thresholds were based partially on suggestions from Section 8.1.4 in the TransCelerate BioPharma Position Paper, and are intended to illustrate some of the flexibility in threshold definitions that are available to the user [1]. The **Default Risk Threshold** data set serves two purposes. First, it gives users the opportunity to start analyzing their data quickly. Second, it assists users in developing their own risk threshold data sets by serving as an initial template, and by providing guidance for how to develop risk thresholds by comparing the data set to the examples described in Section 2.2.3. While the **Default Risk Threshold** data set is certainly useful for getting started, it is not expected to be appropriate across all situations. For example, based on the trial population, certain studies (e.g., those in pediatrics) may require more stringent risk thresholds to alert the sponsor to any important differences in safety across the study sites. Further, trials with one or more adaptations may have more exacting criteria for certain quality measures to ensure that data used for study modifications are as close to error-free as possible. Even over the course of the same trial, risks for certain variables may change based on the comfort level of important procedures among the clinical sites, or the discovery that certain parts of the protocol may have been misinterpreted, necessitating additional training.

Define Risk Threshold Data Set allows the user to select a risk threshold data set to be used as a starting template for creating a new set of thresholds for the RBM analysis and review. The report opens the selected risk threshold data set, permitting the user to change thresholds for moderate or severe risk (yellow or red, respectively) as well as the weights used to combine individual variables into the overall risk indicators (Figure 3.1 on page 81). The same set of risk thresholds are applied to the values of the site- and country-level risk indicator data tables. Risk

3.2 Defining Alternate Risk Thresholds and Actions

thresholds can be specified for as many or as few indicators as desired, and thresholds can be specified for variables that are not available in the current study. In these cases, the thresholds are simply ignored. It is expected that the **Yellow** and **Red Percent of Center**, **Yellow** and **Red Magnitude**, and the **Weight for Overall Risk Indicator** are either missing or ≥ 0. The **Yellow** and **Red Percent** variables determine risk by considering distance from a center value, while the magnitudes indicate minimally meaningful values in order to generate a signal. The **Center Flag** defines the value around which risk is measured, either the Mean or Median value for all sites or countries (depending on the analysis), or a **Fixed** value set by the user in **Center Value**. **Direction for Risk Signals** specifies whether excess beyond the **Center Flag** only is considered risk (U for Upper), scarcity below the **Center Flag** only is considered risk (L for Lower), or risk is measured in Both (= B) directions.

Figure 3.1 Define Risk Threshold Data Set

#	Variable	Label	Category	Yellow Percent of Center	Yellow Magnitude	Red Percent of Center	Red Magnitude	Weight for Overall Risk Indicator	Center Flag	Center Value	Direction for Risk Signals
1	TOTAL	Total Subjects	Enrollment	.	.	.	.	.	.	.	.
2	RAND	Randomized	Enrollment	.	.	.	.	.	.	.	.
3	SCRNFAIL	Screen Failure	Enrollment	.	.	.	.	.	.	.	.
4	PCTSCRNFAIL	Percent Screen Fail of Total Subjects	Enrollment	5	0	15	0	1	Median	.	B
5	TREATED	Treated	Enrollment	.	.	.	.	.	.	.	.
6	PATWEEKS	PatientWeeks on Study	Enrollment	.	.	.	.	.	.	.	.
7	MISSCONSENT	Missing Informed Consent	Enrollment	.	.	.	1	1	Fixed	0	U
8	WEEKSACTIVE	Weeks Active	Enrollment	.	.	.	.	.	.	.	.
9	SITERATE	Randomized per Week Active	Enrollment	5	0	15	0	1	Median	.	U
10	EXPRAND	Expected Randomized	Enrollment	.	.	.	.	.	.	.	.
11	DIFFRAND	Observed Minus Expected Randomized	Enrollment	.	5	.	10	.	.	.	L
12	TARGETRAND	Target Randomized	Enrollment	.	.	.	.	.	.	.	.
13	TARGETDIFFRAND	Observed Minus Target Randomized	Enrollment	.	5	.	10	.	.	.	L
14	COMPLETE	Completed	Disposition	.	.	.	.	.	.	.	.
15	PCTCOMP	Percent Completed of Randomized Subjects	Disposition	.	.	.	.	.	.	.	.
16	ONGOING	Ongoing	Disposition	.	.	.	.	.	.	.	.
17	PCTONGOING	Percent Ongoing of Randomized Subjects	Disposition	.	.	.	.	.	.	.	.
18	DISCON	Discontinued	Disposition	.	.	.	.	.	.	.	.
19	PCTDISCON	Percent Discontinued of Randomized Subjects	Disposition	15	3	30	3	1	Median	.	B
20	DISCDEATH	Discontinued due to Death	Disposition	.	.	.	.	.	.	.	.
21	DCAE	Discontinued due to AE	Disposition	.	.	.	.	.	.	.	.
22	LTFU	Lost to Followup	Disposition	.	.	.	.	.	.	.	.
23	WITHDREW	Patient Withdrew from Study	Disposition	.	.	.	.	.	.	.	.
24	DCOTH	Discontinued Other	Disposition	.	.	.	.	.	.	.	.
25	TOTIE	Total Inclusion or Exclusion Not Met	Disposition	.	.	.	.	.	.	.	.
26	AVGIE	Average Inclusion or Exclusion Not Met per Randomize	Disposition	5	0	15	0	1	Median	.	B
27	TOTDV	Total Protocol Deviations	Disposition	.	.	.	.	.	.	.	.
28	AVGDV	Average Deviations per Randomized Subject	Disposition	5	0	15	0	1	Median	.	B
29	DEATH	Death	Adverse Events	.	.	.	.	.	.	.	.
30	PCTDEATH	Percent Deaths of Randomized Subjects	Adverse Events	5	0	10	0	0	Median	.	U
31	PWDEATH	Deaths per PatientWeek	Adverse Events	5	0	10	0	1	Median	.	U
32	TOTAE	Total AEs on Study	Adverse Events	.	.	.	.	.	.	.	.
33	AVGAE	Average AEs per Randomized Subject	Adverse Events	5	0	15	0	0	Median	.	B
34	PWAE	AEs per PatientWeek	Adverse Events	5	0	15	0	1	Median	.	B
35	TOTSAE	Total SAEs on Study	Adverse Events	.	.	.	.	.	.	.	.
36	AVGSAE	Average SAEs per Randomized Subject	Adverse Events	5	0	15	0	0	Median	.	B
37	PWSAE	SAEs per PatientWeek	Adverse Events	5	0	15	0	1	Median	.	B
38	TOTALQUERIES	Total Queries	Manually Entered	.	.	.	.	.	.	.	.
39	AVTOTQ	Average Queries per Randomized Subject	Manually Entered	5	0	15	0	0	Median	.	U
40	PWTOTQ	Queries per PatientWeek	Manually Entered	5	2	15	2	1	Median	.	U
41	OVERDUEQUERIES	Overdue Queries	Manually Entered	.	.	.	.	.	.	.	.
42	AVOVERQ	Average Overdue Queries per Randomized Subject	Manually Entered	5	0	15	0	0	Median	.	U
43	PWOVERQ	Overdue Queries per PatientWeek	Manually Entered	5	2	15	2	1	Median	.	U

Risk thresholds work the same for individual indicators as they do for Overall Indicators. In general, for the *i*th site or country and the *j*th risk indicator:

1. moderate risk (yellow) occurs when **Yellow Percent of Center** < 100 × c_{ij} / μ_j ≤ **Red Percent of Center** with x_{ij} ≥ **Yellow Magnitude**, and

2. severe risk (red) occurs when 100 × c_{ij} / μ_j > **Red Percent of Center** with x_{ij} ≥ **Red Magnitude**,

where μ_j is the Mean, Median or user-supplied **Center Value**, and x_{ij} is the value for the *i*th site or country and the *j*th risk indicator. The quantity c_{ij} equals $\left|x_{ij} - \mu_j\right|$, $\left(x_{ij} - \mu_j\right)$ and $-\left(x_{ij} - \mu_j\right)$, for **Direction for Risk Signals** equal to B, U, and L, respectively. It is acceptable to specify both yellow and red risk thresholds, or one or no risk thresholds. When specifying only a moderate threshold, the **Red Percent of Center** is left missing in the risk threshold data set so that moderate risk is considered $100 \times c_{ij} / \mu_j >$ **Yellow Percent of Center**. In instances where values do not meet the criteria for moderate or severe risk, the risk is considered mild (green). Note that for risk thresholds defined using the above criteria, no threshold colors are determined in instances where the Mean, Median, or **Center Value** is calculated or set to zero.

For example, from Table 2.5 on page 44 2.5 moderate risk for **Percent Discontinued of Randomized Subjects** is defined as >15% and ≤30% more or less than median value of all sites with an observed percent discontinued of at least 3; severe risk is defined as >30% more or less than median value of all sites with an observed percent discontinued of at least 3. Compare this to the entries in the data table in Figure 3.2 on page 82. Here, the median discontinuation percentage for all sites or countries is used as a benchmark to compare the individual values. Further, since I am interested in considering extreme values above and below the median as risky, **Direction for Risk Signals** is specified as B. There is also the additional caveat that the value must be at least 3 for me to elevate risk to moderate or severe. In other words, there must be at least 3% of subjects discontinuing from the site (or country) and considered "extreme" compared to the median to elevate risk. If I was interested in sites only with excessive discontinuations, I would change the **Direction for Risk Signals** to U. Or if I expect discontinuations to be extremely rare, I may choose to measure risk from a fixed **Center Value** of 0. Alternatively, it may be more appropriate to define risks thresholds based only upon magnitudes for outcomes that aren't likely to occur very frequently; see the later example for missing informed consent.

Figure 3.2 Risk Thresholds for Percent Discontinued of Randomized Subjects

Label	Category	Yellow Percent of Center	Yellow Magnitude	Red Percent of Center	Red Magnitude	Weight for Overall Risk Indicator	Center Flag	Center Value	Direction for Risk Signals
Percent Discontinued of Randomized Subjects	Disposition	15	3	30	3	1	Median		B

Risk thresholds can also be defined based solely only the magnitudes of the values observed. In this case,

1 moderate risk (yellow) occurs when **Yellow Magnitude** $\leq c_{ij} <$ **Red Magnitude**, or

2 severe risk (red) occurs when $c_{ij} \geq$ **Red Magnitude**,

where the quantity c_{ij} is defined as above. Here, it is acceptable to specify both, one, or no risk thresholds. When specifying only a moderate threshold, the **Red Magnitude** is left missing in the risk threshold data set so that moderate risk is $c_{ij} \geq$ **Yellow Magnitude**. In cases where moderate or severe risk does not apply, the risk is considered mild (green).

Consider Table 2.5 on page 44 again. Moderate risk for **Observed Minus Expected Randomized** is defined as ≥5 and <10 subjects below the median of expected enrollment using **Expected Randomized**, which is based on the average performance over all sites or countries; severe risk is defined as ≥10 subjects below the median of expected enrollment. Compare this to the entries in the data table in Figure 3.3 on page 83. Here, we are only interested in underperforming sites so that **Direction for Risk Signals** is equal to L. However, it may be of interest to change this value to a B to examine sites (or countries) that are extreme in their over-performance; this may indicate a laxness in assessing the eligibility criteria for the clinical trial. Notice that **Weight for Overall Risk Indicator** is missing. While zero is the preferred value, the interpretation is this: While I choose to calculate risks for this individual variable, I do not wish to include it in any of the overall risk indicators.

Figure 3.3 Risk Thresholds for Observed Minus Expected Randomized

Label	Category	Yellow Percent of Center	Yellow Magnitude	Red Percent of Center	Red Magnitude	Weight for Overall Risk Indicator	Center Flag	Center Value	Direction for Risk Signals
Observed Minus Expected Randomized	Enrollment	.	5	.	10	.		.	L

For one final example, **Missing Informed Consent** only has a severe risk category defined. If there is a single instance of a missing informed consent form, the site or country is flagged with a severe risk (Figure 3.4 on page 83). In general, getting consent from the patient is the first step prior to performing any study procedures; failure to gain written consent is a serious breach of GCP. Here, **Center Flag** is set to Fixed with a **Center Value** of 0 to indicate that I would like to calculate the risk contribution from 0 in lieu of the mean or median (which would reduce risk somewhat), since my expectation is that this event should never occur within the study. Fixed center values (of 0, say) could also be applied toward other variables such as queries or protocol deviations. In this case, I would consider any instance of these occurring as elevating the risk for the particular site or country. In general, however, queries and deviation are expected to occur, so measuring risk from some overall average (mean or median) is perhaps sufficient. However, if there is a certain target value, say 5 queries per randomized subject (**Average Queries per Randomized Subject**) that is preferable to set as a benchmark from which to measure risk, **Center Flag** is set to Fixed with a **Center Value** of 5.

Figure 3.4 Risk Thresholds for Missing Informed Consent

Label	Category	Yellow Percent of Center	Yellow Magnitude	Red Percent of Center	Red Magnitude	Weight for Overall Risk Indicator	Center Flag	Center Value	Direction for Risk Signals
Missing Informed Consent	Enrollment	.	.	.	1	1	Fixed	0	U

The **Define Risk Threshold Data Set** report has two drill downs available. The first, **Save Risk Threshold Data Set**, allows the user to save any modifications to a new risk threshold data set or update changes to previously defined data sets (Figure 3.5 on page 84).

Figure 3.5 Save Risk Threshold Data Set Drill Down

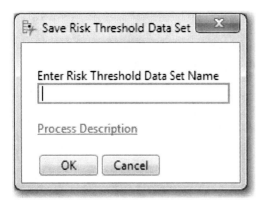

Note that it is not possible to overwrite the **Default Risk Threshold** data set. The second drill down, **Check Risk Threshold Data Set**, generates a PDF report from a brief SAS program to ensure that the values entered into in the data table are appropriately defined. Edit checks performed include:

1 **Yellow** and **Red Magnitude**, when defined, are ≥ 0. If a negative value is presented, the absolute value is used in the analysis.

2 **Yellow** and **Red Percent of Center**, when defined, are ≥ 0. If a negative value is presented, the absolute value is used in the analysis.

3 **Yellow Percent of Center** must be less than **Red Percent of Center**.

4 Checks of missing **Weight for Overall Risk Indicator**. These values are assumed 0 when left missing if risk thresholds are defined.

5 Checks that **Center Flag** and **Direction for Risk Signals** contain appropriate values. If **Center Flag** is specified and **Center Value** is missing, the latter is assumed 0 in the analysis.

Opening the PDF report for the **Default Risk Threshold** data set shows that two weights are left missing for risk indicators (**Observed Minus Expected Randomized** and **Observed Minus Target Randomized**) with defined criteria. These will be assumed 0 by the analysis program. In addition, **Center Flag** is not specified for the same variables; the missing values will be assumed to be Median by the analysis program. The values in any risk threshold data set can be modified until no warnings are presented in the PDF report. Otherwise, the data set can be left as is if the user is comfortable with the assumptions made for the analysis.

Table 3.1 on page 85 describes the contents of the risk threshold data sets.

Table 3.1 Contents of Risk Threshold Data Sets

Variable	Details
Variable	The name of the variable used within the SAS program. Used to merge in risk thresholds. Only of interest to those individuals wanting to modify the underlying SAS program. This includes any user-added variables from **Update Study Risk Data Set**. Each added variable will add a USERVARx, a AVUSERVARx averaging by the number of randomized subjects, and a PWUSERVARx averaging by the number of **PatientWeeks**.
Label	The label of the SAS variable, and the values displayed within all of the JMP output. You can switch back and forth between the names and labels in the JMP data tables using **View > SAS Names** or **View > SAS Labels**. Labels provided for user-added variables from **Update Study Risk Data Set** will include the user-added text for USERVARx, user-added text "per Randomized Subject" for AVUSERVARx, and user-added text "per PatientWeek" for PWUSERVARx.
Category	The grouping category of the variable in **Risk-Based Monitoring**, this column determines which variables are grouped for the overall risk indicators OVERALLA, OVERALLD, OVERALLAE, and OVERALLM. Note that all data added from **Update Study Risk Data Set** is categorized as "Manually Entered."
Yellow Percent of Center	Determines the upper and lower thresholds for green and yellow, respectively. This is the percentage above or below the center value. Only values ≥ 0 are acceptable.
Yellow Magnitude	The magnitude of the observed value must be at least as great to meet the yellow threshold. Only values ≥ 0 are acceptable.

Variable	Details
Red Percent of Center	Determines the upper and lower thresholds for yellow and red, respectively. This is the percentage above or below the center value. Only values ≥ 0 are acceptable.
Red Magnitude	The magnitude of the observed value must be at least as great to meet the red threshold. Only values ≥ 0 are acceptable.
Weight for Overall Risk Indicator	A weight > 0 means a risk indicator will contribute to the Overall Risk Indicator and the appropriate subgroup of Overall Risk Indicator. Only values ≥ 0 are acceptable. By default, all weights are 1, meaning each variable contributes equally to each overall risk indicator. Weights are meaningless for Overall Indicators and can be left missing.
Center Flag	The value from which risk will be measured using **Yellow Percent of Center** and **Red Percent of Center**. Appropriate values are the Mean or Median, which is the summary value of all clinical sites or countries. "Fixed" implies a fixed value, which is supplied in Center Value.
Center Value	When **Center Flag** = Fixed, the value around which risk is measured.
Direction for Risk Signals	Appropriate values are L for Lower, U for Upper, and B for Both. Risk is measured only in excess for U, and only in scarcity or underreporting for L. B counts risk in both directions.
Yellow Recommended Action	This is the suggested action to address moderate risk. Available from the **Report Actions** drill down in **Risk-Based Monitoring**.
Red Recommended Action	This is the suggested action to address severe risk. Available from the **Report Actions** drill down in **Risk-Based Monitoring**.

Variable	Details
Comment	Space to provide comments about the particular thresholds. Has no impact on analysis.

3.2.2 Thresholds for User-Added Risk Indicators

In Section 2.2.1.2 we added the new variable **Computed Eligibility Violations** to the **Study Risk Data Set** using **Update Study Risk Data Set**. In actuality, **Computed Eligibility Violations** was added to a database of user-defined variables that will be added to the **Study Risk Data Set** of every new study added to JMP Clinical for the purposes of RBM analysis and review. As you may have noticed in "2.4 Reviewing Risk Indicators" on page 53, no risk thresholds were applied to **Computed Eligibility Violations**, **Computed Eligibility Violations per Randomized Subject**, or **Computed Eligibility Violations per PatientWeek** since none of the cells for these variables were colored green, yellow, or red to indicate the level of risk for the individual sites or countries. This is due to the fact that the **Default Risk Threshold** data set does not initially contain rows for these newly defined risk indicators, meaning that no risk thresholds are defined. To ensure that newly added risk indicators have thresholds defined to identify safety or quality signals in **Risk-Based Monitoring**, the user must first run **Define Study Risk Data Set** to add the new risk indicator rows to the data table.

For example, if you run **Define Study Risk Data Set** and choose **Default Risk Threshold** as the **Risk Threshold Data Set** (initially, it will be your only choice), a data table will be presented with added rows corresponding to **Computed Eligibility Violations** below the rows for the overall risk indicators (Figure 3.6 on page 87). All variables for these rows will be missing (blank). Clicking **Save Risk Threshold Data Set** and providing a name at this point (say, **New Default**) will add the variables to a new risk threshold data set that can be selected from drop-down menus in the dialogs, but because thresholds are missing for the **Computed Eligibility Violations** variables, they will still not be evaluated for risk nor added to any overall risk indicators.

Figure 3.6 Undefined Risk Thresholds for Computed Eligibility Violations Indicators

Label	Category	Yellow Percent of Center	Yellow Magnitude	Red Percent of Center	Red Magnitude	Weight for Overall Risk Indicator	Center Flag	Center Value	Direction for Risk Signals
Computed Eligibility Violations	Manually Entered	•	•	•	•	•	•		•
Computed Eligibility Violations per Randomized Subject	Manually Entered	•	•	•	•	•		•	
Computed Eligibility Violations per PatientWeek	Manually Entered	•	•	•	•	•		•	

We can add any risk limits we choose, but let's make things easy by copying and pasting the values from **Computed Deviations**, **Computed Deviations per Randomized Subject**, and **Computed Deviations per PatientWeek** into the **Computed Eligibility Violations** rows (Figure 3.7 on page 88). Click **Save Risk Threshold Data Set**. Based on these definitions, no thresholds are applied to **Computed Eligibility Violations**, and **Computed Eligibility Violations per PatientWeek** will be incorporated into the **Overall Risk Indicator** and **Overall Risk**

Indicator Manually Entered while **Computed Eligibility Violations per Randomized Subject** will not.

Figure 3.7 *Risk Thresholds for Computed Eligibility Violations Indicators*

Label	Category	Yellow Percent of Center	Yellow Magnitude	Red Percent of Center	Red Magnitude	Weight for Overall Risk Indicator	Center Flag	Center Value	Direction for Risk Signals
Computed Eligibility Violations	Manually Entered	.	.	.	.	.		.	
Computed Eligibility Violations per Randomized Subject	Manually Entered	5	0	15	0	0	Median		U
Computed Eligibility Violations per PatientWeek	Manually Entered	5	0	15	0	1	Median		U

Running **Risk-Based Monitoring** with **New Default** selected as the **Risk Threshold Data Set** will generate the following updated columns in the analysis (Figure 3.8 on page 89, uses **Cols > Reorder Columns**). Compare to Figure 2.15, and you will see slight differences in the two **Overall Risk Indicators** when accounting for **Computed Eligibility Violations per PatientWeek**.

Figure 3.8 New Risk Thresholds and Modified Overall Risk Indicators

	Study Site Identifier	Country	Overall Risk Indicator	Overall Risk Indicator Manually Entered	Computed Eligibility Violations	Computed Eligibility Violations per Randomized Subject	Computed Eligibility Violations per PatientWeek
1	01	USA	0.478	0.517	13	0.25	0.15
2	02	USA	0.237	0.041	15	0.47	0.26
3	03	USA	0.117	0.059	15	0.65	0.40
4	04	FRA	0.482	0.662	11	0.42	0.23
5	05	ITA	1.789	1.643	14	2.80	2.13
6	06	USA	1.406	1.363	12	1.71	1.11
7	07	CHN	1.407	1.733	11	2.20	1.71
8	08	GBR	0.547	0.373	16	0.70	0.39
9	09	CAN	0.278	0.153	17	0.43	0.29
10	10	USA	0.265	0.136	12	0.71	0.42
11	12	DEU	0.452	0.267	14	0.88	0.52
12	14	CAN	0.403	0.000	11	0.15	0.08
13	16	USA	0.611	0.505	12	0.31	0.19
14	17	USA	0.342	0.129	12	0.67	0.38
15	18	JPN	0.489	0.476	16	0.73	0.42
16	19	CAN	0.744	0.790	16	1.78	1.14
17	20	FRA	0.147	0.088	16	0.89	0.53
18	21	ESP	0.374	0.489	13	1.30	0.67
19	22	CAN	0.469	0.326	14	0.61	0.34
20	23	USA	0.160	0.161	13	0.45	0.29
21	24	CHN	0.224	0.202	12	0.57	0.36
22	25	USA	0.519	0.405	17	1.13	0.66
23	26	DEU	0.627	0.801	13	1.63	0.92
24	27	CHE	0.186	0.171	13	0.54	0.30
25	28	CAN	0.424	0.280	14	0.19	0.10
26	29	USA	0.608	0.588	24	2.00	1.08
27	30	USA	0.925	1.199	19	3.80	2.25
28	31	ITA	2.178	2.120	13	2.60	3.03
29	32	USA	0.175	0.006	15	0.33	0.19
30	33	USA	0.481	0.788	14	1.56	1.04
31	34	CHN	0.649	0.632	14	2.00	1.21
32	35	USA	0.371	0.529	14	1.75	0.98
33	36	USA	0.709	0.743	19	3.17	1.64
34	37	ESP	0.186	0.197	24	1.85	1.04
35	39	USA	0.316	0.317	16	0.94	0.69
36	40	GBR	0.494	0.468	10	0.26	0.15
37	42	GBR	0.266	0.382	19	0.95	0.57
38	44	USA	0.302	0.223	14	0.37	0.22
39	45	JPN	0.502	0.301	23	1.10	0.69
40	46	USA	0.401	0.317	9	0.28	0.16

3.2.3 Weights for Overall Risk Indicators

In Section 2.2.2, I introduced the five overall indicators that are weighted averages or combinations of the individual risk indicators for which at least one risk threshold is defined, where **Weight for Overall Risk Indicator** > 0, and the standard deviation of the indicator > 0. The **Overall Risk Indicator** incorporates all of the variables meeting these criteria into a single measure that signifies the overall risk and performance of a clinical site. However, if no indicators have a **Weight for Overall Risk Indicator** > 0, then the corresponding **Overall Risk Indicator** is not generated. Each of the other four overall indicators—Enrollment Metrics, Disposition, Adverse Events, and Manually Entered—combines subsets of the risk indicators based on **Category** in the risk weight data set. By default, **Category** matches how variables are grouped in **Risk-Based Monitoring**, with **Manually Entered** applied to all user-supplied risk indicators from **Update Study Risk Data Set**. For a given category, if no indicators have a Weight for Overall Risk Indicator > 0, then the corresponding overall indicator is not provided.

As mentioned earlier, **Weight for Overall Risk Indicator** (w_j) can either be missing (which is assumed 0) or ≥ 0. The weights are self-normalizing, in that each weight is divided by the sum of all weights for variables contributing to the particular overall indicator. The contribution of each indicator to an overall indicator is based upon its weight, center value (either Mean, Median, or user-provided **Center Value**), standard deviation, and **Direction for Risk Signals**. In general, the value for an overall indicator for the ith site or country and the jth risk indicator is defined as

$$\frac{\sum_j w_j c_{ij}}{\sum_j w_j}, \text{ where } c_{ij} = \frac{|x_{ij} - \mu_j|}{\sigma_j}, \; c_{ij} = \frac{\max(0, x_{ij} - \mu_j)}{\sigma_j}, \; c_{ij} = \frac{\max(0, \mu_j - x_{ij})}{\sigma_j} \text{ when Direction for}$$

Risk Signals equals B, U, or L, respectively. This can be interpreted as "larger values imply greater risk." By default, all weights are assumed equal to 1 in the **Default Risk Threshold** data set, meaning that each variable contributes equally to each overall indicator.

So why change the contributions of the individual risk indicators to the overall risk indicators? Simply put, not all data are equal; or rather, some data are more equal than others. The importance some indicators play over others may depend on the particular study design or other characteristics of the clinical trial. For example, a first-in-human study or study in a special population such as pediatrics will likely place greater emphasis on safety than on quality indicators. Though in most clinical trials, it may be appropriate to apply greater weight to more severe safety outcomes. For example, the weights 1, 2, and 3 could be applied to **AEs per PatientWeek**, **SAEs per PatientWeek**, and **Deaths per PatientWeek**, respectively (Figure 3.9 on page 91, using **Rows > Move Rows**). As another example, I could modify the weights for the **Manually Entered** risk indicators so that only **Computed Deviations per PatientWeek** and **Computed Eligibility Violations per PatientWeek** are averaged in the **Overall Risk Indicator Manually Entered**. Click **Save Risk Threshold Data Set** and save as **Adjusted Overall Weights**.

Run **Risk-Based Monitoring** with **Adjusted Overall Weights** as the **Risk Threshold Data Set**. Compare the updated **Overall Risk Indicator Adverse Event** and **Overall Risk Indicator Manually Entered** in Figure 3.10 on page 92 to those contained in Figure 2.15.

Figure 3.9 Modified Overall Risk Indicator Weights for Adverse Events and Manually Entered

	Variable	Label	Category	Yellow Percent of Center	Yellow Magnitude	Red Percent of Center	Red Magnitude	Weight for Overall Risk Indicator	Center Flag	Center Value	Direction for Risk Signals
1	PWDEATH	Deaths per PatientWeek	Adverse Events	5	0	10	0	3	Median		U
2	PWSAE	SAEs per PatientWeek	Adverse Events	5	0	15	0	2	Median		B
3	PWAE	AEs per PatientWeek	Adverse Events	5	0	15	0	1	Median		B
4	PWUSERVAR1	Computed Eligibility Violations per PatientWeek	Manually Entered	5	0	15	0	1	Median		U
5	PWCOMPDEV	Computed Deviations per PatientWeek	Manually Entered	5	0	15	0	1	Median		U
6	PWTOTQ	Queries per PatientWeek	Manually Entered	5	2	15	2	0	Median		U
7	PWOVERQ	Overdue Queries per PatientWeek	Manually Entered	5	2	15	2	0	Median		U
8	PWMISSPAGE	Missing Pages per PatientWeek	Manually Entered	5	0	15	0	0	Median		U
9	QUERYRESP	Query Response Time	Manually Entered	5	0	15	0	0	Median		U
10	CRFENTRYRESP	CRF Entry Response Time	Manually Entered	5	0	15	0	0	Median		U

Figure 3.10 Modified Overall Risk Indicators for Adverse Events and Manually Entered

	Study Site Identifier	Country	Overall Risk Indicator Adverse Events	Overall Risk Indicator Manually Entered
1	01	USA	0.295	0.000
2	02	USA	0.478	0.000
3	03	USA	0.175	0.129
4	04	FRA	0.649	0.000
5	05	ITA	1.669	2.827
6	06	USA	2.585	1.870
7	07	CHN	1.302	2.051
8	08	GBR	1.013	0.001
9	09	CAN	0.428	0.000
10	10	USA	0.775	0.172
11	12	DEU	0.965	0.039
12	14	CAN	0.371	0.000
13	16	USA	0.462	0.000
14	17	USA	0.824	0.020
15	18	JPN	0.376	0.366
16	19	CAN	0.218	1.037
17	20	FRA	0.351	0.045
18	21	ESP	0.330	0.764
19	22	CAN	1.320	0.000
20	23	USA	0.166	0.000
21	24	CHN	0.167	0.000
22	25	USA	0.822	0.384
23	26	DEU	0.450	0.339
24	27	CHE	0.141	0.000
25	28	CAN	0.069	0.000
26	29	USA	1.156	0.945
27	30	USA	0.510	2.488
28	31	ITA	2.703	2.301
29	32	USA	0.145	0.000
30	33	USA	0.307	0.761
31	34	CHN	1.132	1.003
32	35	USA	0.112	0.385
33	36	USA	0.588	2.166
34	37	ESP	0.307	0.461
35	39	USA	0.642	0.331
36	40	GBR	0.212	0.000
37	42	GBR	0.049	0.075
38	44	USA	0.130	0.000
39	45	JPN	0.538	0.166
40	46	USA	0.190	0.011

Finally, while it currently isn't possible to switch the categories (Enrollment Metrics, Disposition, Adverse Events, and Manually-Entered) for variables within the **Site-** or **Country-Level Risk Indicators** data tables, it is possible to choose which subgroup overall risk indicator a variable belongs to by switching the **Category** in a risk threshold data set. For example, because patients often discontinue clinical trials due to some safety or tolerability issue, it may make sense to incorporate **Percent Discontinued of Randomized Subjects** into the **Overall Risk Indicator**

Adverse Event. Run **Define Risk Threshold Data Set** using **Default Risk Threshold** as the **Risk Threshold Data Set**. Change **Category** for **Percent Discontinued of Randomized Subjects** to **Adverse Events**, and Click **Save Risk Threshold Data Set** to save as **Overall AE** (Figure 3.11 on page 93, using **Rows > Move Rows**).

Figure 3.11 Adding Percent Discontinued of Randomized Subjects to Overall Indicator Adverse Events

	Variable	Label	Category	Yellow Percent of Center	Yellow Magnitude	Red Percent of Center	Red Magnitude	Weight for Overall Risk Indicator	Center Flag	Center Value	Direction for Risk Signals
1	PCTDISCON	Percent Discontinued of Randomized Subjects	Adverse Events	15	3	30	3	1	Median	·	B
2	DEATH	Death	Adverse Events	·	·	·	·	·	·	·	·
3	PCTDEATH	Percent Deaths of Randomized Subjects	Adverse Events	5	0	10	0	0	Median	·	U
4	PWDEATH	Deaths per PatientWeek	Adverse Events	5	0	10	0	1	Median	·	U
5	TOTAE	Total AEs on Study	Adverse Events	·	·	·	·	·	·	·	·
6	AVGAE	Average AEs per Randomized Subject	Adverse Events	5	0	15	0	0	Median	·	B
7	PWAE	AEs per PatientWeek	Adverse Events	5	0	15	0	1	Median	·	B
8	TOTSAE	Total SAEs on Study	Adverse Events	·	·	·	·	·	·	·	·
9	AVGSAE	Average SAEs per Randomized Subject	Adverse Events	5	0	15	0	0	Median	·	B
10	PWSAE	SAEs per PatientWeek	Adverse Events	5	0	15	0	1	Median	·	B

Run **Risk-Based Monitoring** with **Overall AE** as the **Risk Threshold Data Set**. Compare the updated **Overall Risk Indicator Adverse Event** in Figure 3.12 on page 94 (uses **Cols > Reorder Columns**) to the one contained in Figure 2.15.

As a final exercise, think about how you could modify **Category** in the **Risk Threshold Data Set** to reproduce the **Overall Risk Indicator** using one of the other overall variables.

3.2.4 Defining Actions for Elevated Risk

Most actions in the **Default Risk Threshold** data set are limited to the overall risk indicators (Table 2.6 on page 48 2.6). In order to modify current or define additional actions, the analyst must run **Define Risk Threshold Data Set**, add or modify criteria for **Yellow Recommended Action** and/or **Red Recommended Action**, and save the risk threshold data set with a new name. Just as no default risk thresholds are provided for user-defined indicators added using **Add Variable** in **Update Study Risk Data Set**, no recommended actions are provided either.

Figure 3.12 Overall Risk Indicator Adverse Events Including Percent Discontinued of Randomized Subjects

	Study Site Identifier	Country	Overall Risk Indicator Adverse Events
1	01	USA	0.345
2	02	USA	0.590
3	03	USA	0.278
4	04	FRA	0.408
5	05	ITA	1.881
6	06	USA	2.183
7	07	CHN	1.538
8	08	GBR	1.125
9	09	CAN	0.457
10	10	USA	0.623
11	12	DEU	0.998
12	14	CAN	0.626
13	16	USA	0.768
14	17	USA	0.885
15	18	JPN	0.700
16	19	CAN	0.446
17	20	FRA	0.310
18	21	ESP	0.358
19	22	CAN	0.921
20	23	USA	0.126
21	24	CHN	0.375
22	25	USA	0.979
23	26	DEU	0.637
24	27	CHE	0.216
25	28	CAN	0.222
26	29	USA	0.946
27	30	USA	0.910
28	31	ITA	2.310
29	32	USA	0.197
30	33	USA	0.184
31	34	CHN	1.005
32	35	USA	0.279
33	36	USA	1.003
34	37	ESP	0.260
35	39	USA	0.404
36	40	GBR	0.298
37	42	GBR	0.173
38	44	USA	0.141
39	45	JPN	0.660
40	46	USA	0.312

3.3 Performing Additional Statistical and Graphical Analyses

3.3.1 Statistical Analyses

3.3.1.1 Hierarchical Clustering of Sites

Since I've described the five report buttons under the **Risk-Based Monitoring** menu, and addressed all of the drill-down buttons within each report, some readers may incorrectly conclude that there are no additional features or functionality available to them. Nothing could be further from the truth. The country-, site-, and even the subject-level risk indicator data tables are of a format (each row is an independent observation) that make them employable by many if not most JMP platforms under the **Analyze** and **Graph** menus, as well as the JMP Clinical reports under **Clinical > Pattern Discovery** and **Clinical > Predictive Modeling**. In fact, I illustrated this flexibility in Section 2.4.1.2, when I used **Analyze > Distribution** to generate histograms for risk indicators that are not be represented in the distribution summary tabs. Here, I'll describe a few more examples to inspire analysts to take their RBM analyses and reviews further.

Upon reviewing the **Site-Level Risk Indicator** data table (Figure 2.15), a natural question arises: Which sites are similar to one another in terms of their performance? This information can be useful to identify the set of sites to utilize in future trials. More specifically, it can help distinguish which sites to approach first based upon how they are grouped according to their risk indicators and the characteristics of the upcoming study. For example, is safety of primary concern in the upcoming trial because of a particularly at-risk study population? This could suggest to the sponsor to select from those sites that neither over- or under-report adverse events or other safety issues. Further, grouping sites according to their performance can be useful in order to develop and apply training or other intervention strategies in an efficient and cost-effective manner. Given the multitude of criteria available, manually determining appropriate clusters of sites can be a challenging task. Here, we'll group sites using the hierarchical clustering methods available under **Analyze > Multivariate Methods > Cluster**.

Run the **Risk-Based Monitoring** report for **Nicardipine** using the **Default Risk Threshold** data set, and click anywhere within the **Site-Level Risk Indicator** data table to make it the current data table. Click on **Overall Risk Indicator Adverse Events** in the **Risk Indicator** drill down. Go to **Analyze > Multivariate Methods > Cluster** (Figure 3.13 on page 96). For this analysis we'll cluster clinical sites based on **Deaths per PatientWeek**, **AEs per PatientWeek**, **SAEs per PatientWeek**, and **Percent Discontinued of Randomized Subjects** to get a better understanding of how sites are related based upon several important safety indicators. I've included **Percent Discontinued of Randomized Subjects** since patients often discontinue clinical trials for safety or tolerability reasons. If I wanted to get more specific and limit discontinuation reasons to those specifically due to death or AE, prior to opening the **Clustering** dialog I could calculate new columns in the risk indicator data table to get specific discontinuation percentages in terms of the number of randomized subjects. I'll show an example of how to do this in "3.4.1 New Discontinuation Variables and Risk Thresholds" on page 108. Be sure to choose **Study Site Identifier** as the **Label**. I'll perform the analysis using the default **Ward** method; see the JMP documentation on Hierarchical Clustering for descriptions of the other methods. Click **OK**.

Figure 3.13 Clustering Dialog

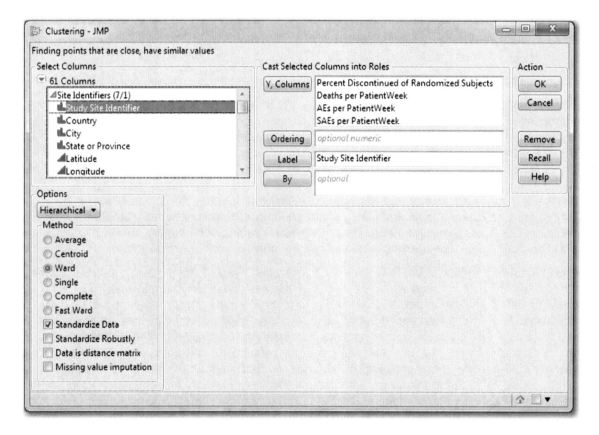

The results of the hierarchical clustering are presented in Figure 3.14 on page 97. The diamonds above and below the dendrogram constitute the **NCluster Handle** (an option in the red triangle menu), which can be dragged left or right to select a number of clusters for the sites (currently at 5 clusters). Choosing **Color Clusters** under the red triangle menu will color site markers according to the selected clustering. Clicking on lines in the dendrogram will highlight the sites from the selected cluster. The markers are currently colored according on the **Overall Risk Indicator Adverse Events** in the **Risk Indicator** drill down. In general, sites are clustered fairly well according to the color of this overall indicator; other variables can be chosen from the **Risk Indicator** drill down to change the color of the cluster markers. Numerous options are available from the red triangle menu for identifying the appropriate number of clusters, alternate plots, and ways to focus on various parts of the dendrogram. The user should review **Help > Books > Multivariate Methods** for more details on the various options available for hierarchical clustering.

Figure 3.14 Hierarchical Clustering of Sites Using Safety Indicators

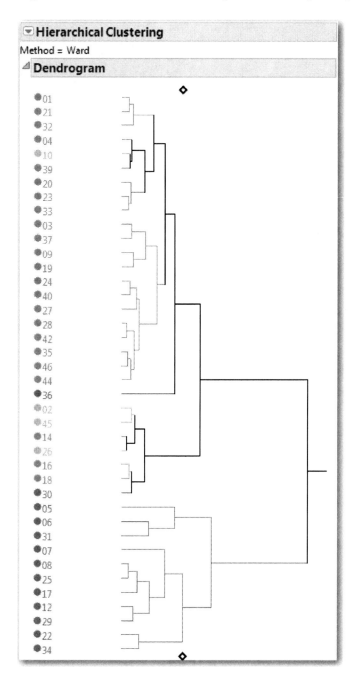

Figure 3.15 Two Way Clustering of Sites and Safety Risk Indicators

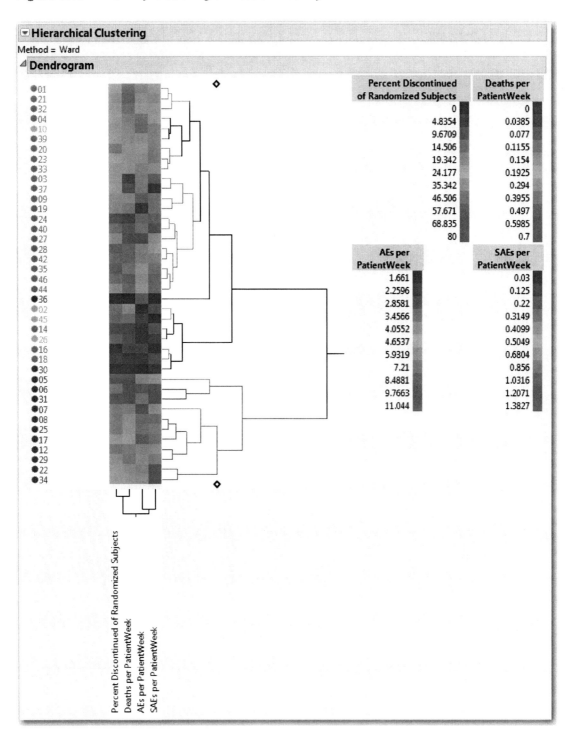

One option that is particularly useful to identify differences among the covariates used within the analysis is available under the red triangle menu as **Two Way Clustering** (Figure 3.15 on page 98). This option provides not only a dendrogram clustering the four covariates used within the analysis, but also a heat map illustrating the magnitudes of the covariates for each site. The **Legend** is provided by selecting the appropriate option under the red triangle menu. The legend is not provided by default since many covariates would generate a very large legend; often the color patterns among the covariates are enough to distinguish how sites vary from one another. Alternatively, choosing not to standardize data will produce a single legend with a unified color scheme. This may be acceptable for risk indicators that are on a similar scale. Here, we obtain some insight as to why site 36 may not be grouped with the other severe centers. This site has only one SAE resulting in a very small **SAE per PatientWeek** value and no discontinuations. While our criteria for the **Overall Risk Indicator Adverse Events** may have considered this site severe, based on the clustering analysis we may choose to handle this site a bit differently if any intervention is to be applied to the clinical sites based on our analysis findings. Of course, similar cluster analyses can be performed at the country level to assess any similarities among countries. Since results may change based on the selected method, users may want to assess the sensitivity of findings to other clustering approaches. Users preferring to use k-means clustering should choose the option **KMeans** from the **Clustering** dialog.

Figure 3.16 JMP Clinical Pattern Discovery Menu

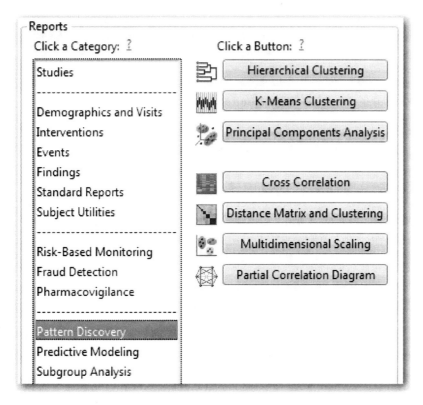

Interested users may wish to take advantage of the **Pattern Discovery** menu from JMP Clinical (Figure 3.16 on page 99). Though some analyses such as **Hierarchical Clustering**, **K-Means**

Clustering, and **Principal Components Analysis** are available in the **Analyze** menu, additional options and drill downs may be surfaced to the user. Details are available in **Help > Books > JMP Life Sciences User Guide** or users can run example settings by clicking **Settings** within the dialogs to display any available examples. Other analyses such as **Multidimensional Scaling** and **Partial Correlation Diagram** are available only from the **Pattern Discovery** menu. In order to use these dialogs, however, the appropriate SAS data sets must be supplied. To determine the SAS data set names of the RBM data sets currently open using **Risk-Based Monitoring**, select **View > Window List**. This window informs the user of all data tables currently available from open analyses. The RBM data tables are of the form rbm_site_sum_XXX, rbm_country_sum_XXX, and rbm_subjectsummary_XXX where XXX is a three-digit numeric value used to maintain every analysis performed within JMP Clinical (that is, until the user manually deletes or empties the study output folder using **Studies > Manage Studies > Clear Study Output Folder**). The data sets are available in the **Output Folder**, which can be accessed from the **Output** tab of any JMP Clinical dialog.

3.3.1.2 Identifying Extreme Sites Using Mahalanobis Distance

Risk thresholds are one way to assess site performance for safety and quality concerns. Though we may examine the risk associated with each individual indicator, I discussed in detail in Section 2.5.3 how it is possible to combine all variables or subsets of related variables into overall measures to assess the risk of clinical sites (or the countries in which these sites are located). Risk severity based on the colors green, yellow, and red are important to understand when meaningful boundaries are crossed, but it is difficult to understand just how different sites may be from one another based on the observed values presented in a data table. Therefore, in Section 2.4.1.2 I presented an alternate and complementary approach to compare the risk among sites by examining histograms and box plots to identify any outliers. Here, we are able to assess any large differences in magnitude for a given risk indicator from sites that may be classified within the same risk severity. For example, Figure 2.15 displays four of nine sites with a severe **Overall Risk Indicator** that can be considered numerically extreme when compared to the other sites.

While we can assess whether there are outliers for any of the overall risk indicators, this approach does not really consider whether sites (or countries) are extreme in the multivariate space based on the contributions of the individual covariates. Mahalanobis distance can be used to calculate the distance between each of k observations to the multivariate mean of p covariates while taking into account the correlation among the individual variables [2]. Though I'll describe this statistic in more detail in Section 4.4.1, we'll make use of it here to identify outlier sites based on the totality of their data. Again, let's make use of the **Site-Level Risk Indicator** data table used in "3.3.1.1 Hierarchical Clustering of Sites" on page 95 and select **Analyze > Multivariate Methods > Multivariate**. For this analysis, we'll limit the indicators to those metrics that were provided in **Update Study Risk Data Set**, including the **Computed Eligibility Violations** indicator created in Section 2.2.1.2 (Figure 3.17 on page 101). Click **OK**.

The output of the **Multivariate and Correlations** platform initially summarizes the pairwise correlations of the seven indicators supplied for analysis. The output presents a **Scatterplot Matrix** examining each pair of variables with a 95% bivariate normal density ellipse provided by default. In lieu of the density ellipses, we can fit regression lines by selecting **Fit Line** and deselecting **Density Ellipse** under the **Scatterplot Matrix** red triangle menu to assess correlation between pairs of variables (Figure 3.18 on page 102). Note that points are colored in Figure 3.18

according to the **Overall Risk Indicator Manually Entered**, which is selected in the **Risk Indicators** drill down. In order to calculate Mahalanobis distance, go to the **Multivariate and Correlations** red triangle menu and select **Outlier Analysis > Mahalanobis Distances** (Figure 3.19 on page 103).

Figure 3.17 Multivariate and Correlations Dialog

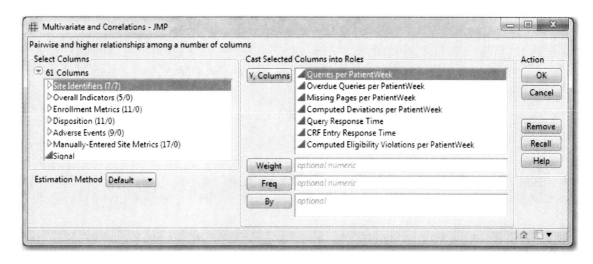

102 Chapter 3 / Risk-Based Monitoring: Customizing the Review Experience

Figure 3.18 Scatterplot Matrix of Manually Entered Risk Indicators

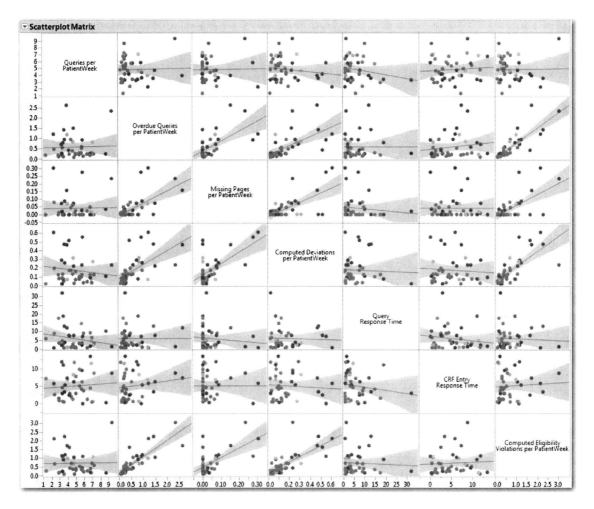

The y-axis is the Mahalanobis distance of each site from the multivariate mean or centroid. Smaller distances indicate sites that are closer to the centroid. Based on the reference line, six sites are considered extreme based on these seven covariates; this is in contrast to the 14 sites with red severity for **Overall Risk Indicator Manually Entered**. Performing hierarchical clustering and generating a heat map as described in Section 2.6.1.1 can provide insight into how individual covariates may contribute to outlier status. Other multivariate outlier analyses are possible from the red triangle menu: **Jacknife Distances** or **T Square**, which provide similar displays to Figure 3.19 on page 103.

Figure 3.19 Mahalanobis Distance for Manually Entered Risk Indicators

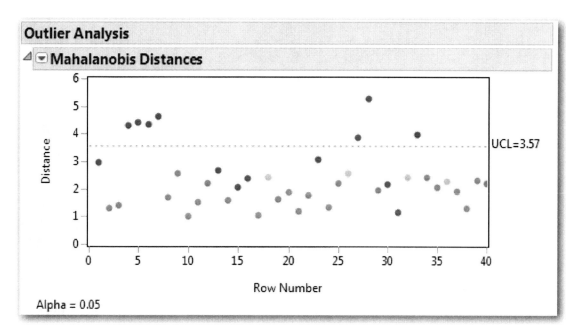

3.3.2 Graphing

3.3.2.1 Bar Charts

Of course, one of the big benefits of JMP is its ability to easily generate interactive and dynamic graphics to help users discover patterns, identify interesting signals, and easily describe their data to a wider audience. In fact, graphics are such an important part of JMP that an entire menu is dedicated to this platform: Graph. In particular, **Graph Builder** is a straightforward drag-and-drop environment to quickly and easily create any number of plots or maps based on the data available in the currently active data table. Here, I'll describe a few examples for plots that may be of interest to users employing the RBM functionality of JMP Clinical. This section also address why more graphical summaries aren't provided within **Risk-Based Monitoring** through drill downs.

Similar to the other analyses described so far in this section, we'll make use of the site-level risk indicator data. Run the **Risk-Based Monitoring** report for **Nicardipine** using the **Default Risk Threshold** data set, and click anywhere within the **Site-Level Risk Indicator** data table to make it the current data table. Then go to **Graph > Graph Builder** to open the **Graph Builder** display (Figure 3.20 on page 104). Additional detail for **Graph Builder** is available in **Help > Books > Essential Graphing**.

Figure 3.20 Graph Builder

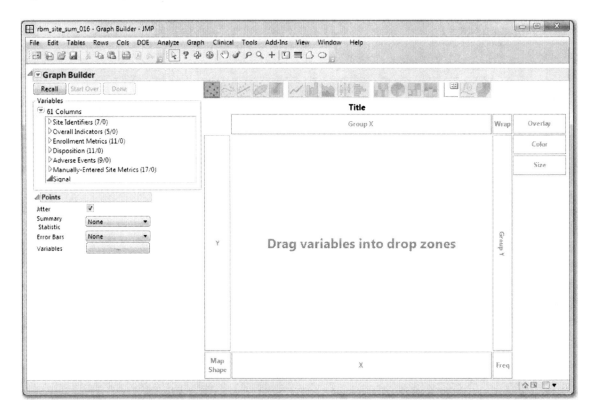

You'll notice that the variables or columns available in the **Site-Level Risk Indicator** data table are presented to the left, organized within their grouping categories; these categories can be expanded by clicking the triangles to the left of the group names. To the right is the **Graph Builder** graph and drop zone. To generate plots, drag variables from the **Columns** list into the appropriate drop zones. Choose plot types by selecting one of the 17 available plot icons from the element type icon toolbar above the **Group X** drop zone. Note that some plots may not be available based on selected variable modeling types (continuous, ordinal, nominal) or the particular drop zones that are utilized. Depending on the data, you can change the modeling type by clicking on the red, green, or blue symbol to the left of each variable and choosing a new type. It will take experience to determine how best to create graphics using **Graph Builder**; however, there are numerous examples in the aforementioned graphing book and throughout the JMP documentation to get you started. Further, you can always click **Start Over** to delete your current attempt and begin again.

Drag **Study Site Identifier** (a nominal variable) to the **X** drop zone, and **Overall Risk Indicator** (continuous) to the Y drop zone. Then click on the **Bar** icon in the element type icon toolbar (Figure 3.21 on page 105). Based on the selected **Summary Statistic**, the y-axis represents the mean. Since the data table has a single row per site, however, a similar plot will be generated for many of the possible summary statistics. Now, your bar chart is finished. Clicking **Done** will close the control panel and tidy up the workspace; the control panel can be opened by clicking on the red triangle menu and selecting **Show Control Panel**. Options to modify various features within the plot are available in the control panel and the red triangle menu, or by right-clicking on either axis or within the main body of the figure. Titles can be modified by double-clicking on the text and providing new information.

Figure 3.21   Bar Chart of Overall Risk Indicator by Study Site Identifier

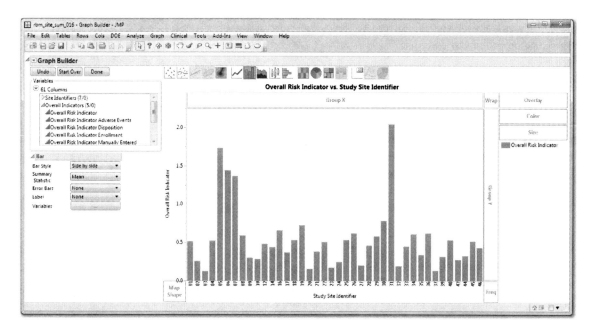

Figure 3.22   Bar Chart of Overall Risk Indicator by Study Site Identifier Grouped by Monitor

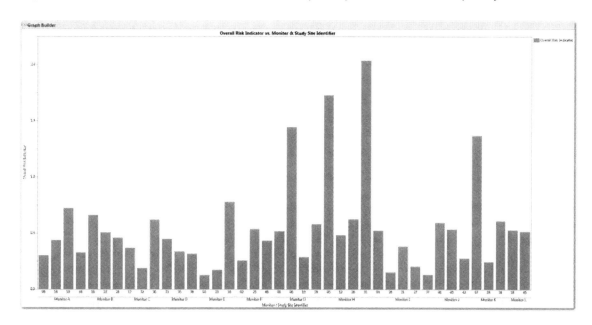

Of course, more informative bar charts can be generated. For example, drag the variable **Monitor** just below **Study Site Identifier** in the **X** drop zone. This will group sites by clinical trial monitor to assess the performance of each individual (Figure 3.22 on page 105). How will the plot change if Monitor is dragged to **Group X** instead?

Of course, with **Graph Builder**, the graphing possibilities are almost limitless. Other available plot types include box plots, scatterplots (with or without regression lines), histograms, heat maps, tree maps, and pie charts. In the next subsection, I illustrate how **Graph Builder** can provide some additional mapping capability of RBM risk indicators.

3.3.2.2 Maps

Maps play a central role in the RBM analysis and review to help identify any potential geographic or country-specific influences on safety or quality. The embedded maps in **Risk-Based Monitoring** are built using **Graph Builder**; the control panel can be surfaced by clicking on the red triangle menu. While it is possible to change map details or modify axes within the report output, I'll illustrate how to generate additional maps from a new instance of **Graph Builder**.

Figure 3.23 Randomized Subjects per Week Active by Study Site Identifier

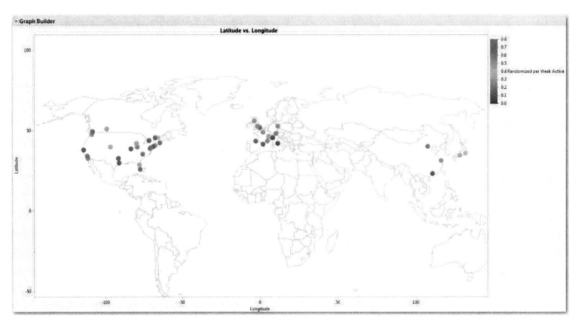

Run the **Risk-Based Monitoring** report for **Nicardipine** using the **Default Risk Threshold** data set, and click anywhere within the **Site-Level Risk Indicator** data table to make it the current data table. Then go to **Graph > Graph Builder** to open the **Graph Builder** display. Drag **Latitude** to **Y** and **Longitude** to **X** to plot the locations of clinical sites. Click on the second icon in the element type icon toolbar to remove the non-parametric **Smoother**. Right-click in the body of the graph and go to **Graph > Background Map > World** to add country boundaries to the map. Markers are currently colored based on the **Overall Risk Indicator** selected in the **Risk Indicators** drill down from the **Risk-Based Monitoring** report. Selecting other variables from this drill down will change the marker color. Drag **Randomized per Week Active** to the **Color** drop zone, and increase marker size by right-clicking in the plot and going to **Graph > Marker Size > XXXL**. This generates a map to summarize site enrollment rate with a legend that spans the range of possible values (Figure 3.23 on page 106). Of course, any risk indicator can be dragged to the **Color** drop zone. In lieu of changing marker size through menus as above, markers could be sized by a second variable in the **Size** drop zone, say **Percent Completed of Randomized Subjects**, to provide additional detail in the map.

Figure 3.24 Randomized Subjects per Week Active by Country

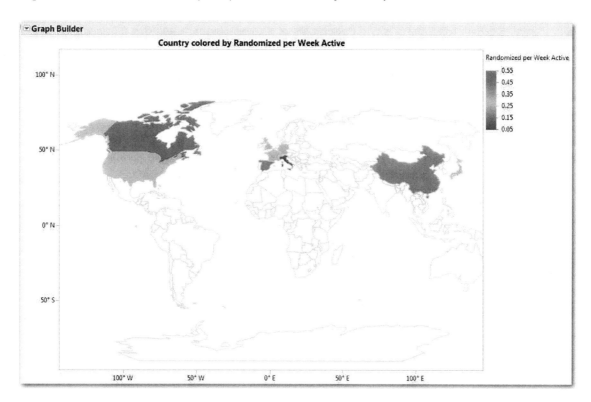

Return to the **Risk-Based Monitoring** report for **Nicardipine** and click anywhere within the **Country-Level Risk Indicator** data table to make it the current data table. Then go to **Graph > Graph Builder** to open the **Graph Builder** display. The first variable in the variable drop down should be **Country**. Drag **Country** to the **Map Shape** drop zone. This will draw currently available countries colored according to the indicator selected in the **Risk Indicator** drill down. Right-click in the figure and go to **Map Shapes > Show Missing Shapes** to draw the remaining countries in the global map. Drag **Randomized per Week Active** to the **Color** drop zone to create Figure 3.24 on page 107. Of course, the user can right-click on the legend to modify any color options, including the gradient, transparency, and color scheme.

3.4 Creating JMP Scripts and Add-Ins

3.4.1 New Discontinuation Variables and Risk Thresholds

My attempt to create reasonably-sized data tables of risk indicators resulted in 61 and 55 variables for the site- and country-level data tables, respectively, for the study **Nicardipine**. Despite this generous set of variables, users continually ask for new and varied ways of summarizing their data for RBM analysis and review. This is to be expected: No two clinical trials are the same, no two study teams are the same, and no two pharmaceutical companies are the same. The functionality in **Update Study Risk Data Set** is one way of generating new risk indicators for data that is typically not captured as part of the study database by averaging by randomized subjects or patient weeks. However, other data from the trial database may be of interest for the RBM analysis and review. For example, the study team may be interested in tracking the number of hospitalizations experienced by patients in the clinical trial and collected in the SDTM domain Healthcare Encounters (HO). For now, except for modifying the underlying SAS code (a topic beyond the scope of this text), there is no means of including indicators based upon other CDISC domains apart from supplying site-level summary statistics through **Update Study Risk Data Set**.

Example Code 3.1   *JSL Script to Add Variables to Risk Indicator Data Tables*

```
Names Default to Here(1);
dt = Current Data Table();

/*
Red: RGB Color({219,146,132})
Yellow: RGB Color({238,238,160})
Green: RGB Color({187,216,189})
*/

name1 = "Percent Discontinued due to Death";
name2 = "Percent Discontinued due to AE";
name3 = "Percent Lost to Followup";

dt << New Column( name1, Format( "Fixed Dec", 8, 1 ),
    Formula( :Discontinued due to Death/:Randomized*100));
dt << New Column( name2, Format( "Fixed Dec", 8, 1 ),
    Formula( :Discontinued due to AE/:Randomized*100));
dt << New Column( name3, Format( "Fixed Dec", 8, 1 ),
    Formula( :Lost to Followup/:Randomized*100));

var1 = Column(dt,name1);
```

```
    var2 = Column(dt,name2);
    var3 = Column(dt,name3);

For Each Row(
    // Looking for percent increases or decreases from the overall mean
    of sites or countries
    If( abs(var1[]-Col Mean(var1))/Col Mean(var1) > 20,
            var1 << Color Cells (RGB Color({219,146,132}), row()),
        10 < abs(var1[]-Col Mean(var1))/Col Mean(var1) <= 20,
            var1 << Color Cells (RGB Color({238,238,160}), row()),
        0 <= abs(var1[]-Col Mean(var1))/Col Mean(var1) <= 10,
            var1 << Color Cells (RGB Color({187,216,189}), row()),
    );
    // Looking for percent increases or decreases from the overall median
      of sites or countries
    If( abs(var2[]-Col Quantile(var2,0.5))/Col Quantile(var2,0.5) > 20,
            var2 << Color Cells (RGB Color({219,146,132}), row()),
        10 < abs(var2[]-Col Quantile(var2,0.5))/Col Quantile(var2,0.5) <= 20,
            var2 << Color Cells (RGB Color({238,238,160}), row()),
        0 <= abs(var2[]-Col Quantile(var2,0.5))/Col Quantile(var2,0.5) <= 10,
            var2 << Color Cells (RGB Color({187,216,189}), row()),
    );
    // Looking for magnitudes of observed percentages that pass certain thresholds
    If( var3[] > 5, var3 << Color Cells (RGB Color({219,146,132}), row()),
        0< var3[] <= 5, var3 << Color Cells (RGB Color({238,238,160}), row()),
        var3[] == 0, var3 << Color Cells (RGB Color({187,216,189}), row()),
    );
);

var1 << Set Selected(1);
dt << Move Selected Columns( After(:Discontinued due to Death));
var1 << Set Selected(0);

var2 << Set Selected(1);
dt << Move Selected Columns( After(:Discontinued due to AE));
var2 << Set Selected(0);

var3 << Set Selected(1);
dt << Move Selected Columns( After(:Lost to Followup));
var3 << Set Selected(0);
```

However, it is possible to define additional variables based on the information naturally collected and summarized in either the **Site-Level Risk Indicator** or **Country-Level Risk Indicator** data tables. Any individual can add new variables to these JMP data tables as functions of the other risk indicators. In this section, we'll take advantage of JMP Add-In functionality to develop a command in the JMP **Add-Ins** menu that will streamline the calculation of new variables each time the RBM analysis and review is updated with new data. For our example, I will calculate discontinuation percentages for several subtypes of discontinuations. Further, I will calculate and apply risk thresholds to color and identify any sites or countries with elevated risk for these new indicators. Since we need to take advantage of JMP scripting, this section is considered an advanced topic. Users may enter the script manually within a script window (Ctrl-T or pushing the **New Script** icon on the toolbar opens a script window), or download **Script to Add Variables.jsl** from the companion website (Example Code 3.1 on page 108). Drag this JSL script onto the JMP Clinical starter menu to open.

The risk indicator data tables provide the percentage of discontinued subjects as a total of randomized subjects. I am interested in calculating three new variables: the **Percent Discontinued due to Death**, the **Percent Discontinued due to AE**, and the **Percent Lost to Followup**. Names Default to Here(1) prevents variables within this script from interacting with variables defined in any other scripts employed by the user or the JMP Clinical system. The Current Data Table() function applies the script to the data table considered to be current (typically, the one that was manipulated last). This is a straightforward way of applying a script to a set of data without specifically referencing the data table using its name. After making the **Site-Level Risk Indicators** current (typically, by interacting with it), run this script to generate the new variables in the site-level data. Next, make the country-level data table current and rerun the script to add variables to the country-level table. For each variable, a new column is added to the data table and a percentage is calculated and formatted to one decimal place. As a note to new JSL scripters, variables are referenced with a colon preceding the variable name to identify them as columns of a JMP table. The For Each Row() function will determine whether each row meets the criteria for signal thresholds and color the cell accordingly. Col Mean() and Col Quantile(, 0.5) calculate the mean or median for the variable, respectively, as the value from which to measure risk. Risk threshold descriptions are provided in Table 3.2 on page 110. The final nine lines of code move the new variables in the data table so that they are located next to their corresponding frequencies. Selecting either of the RBM data tables and pushing the **Run** icon (the red man) in the toolbar will generate the new variables.

Figure 3.25 on page 111 presents the new variables in the site-level data table. Notice that **Percent Discontinued due to AE** was not colored according to the risk thresholds. This is due to the fact that the median value is equal to 0, making calculation of the ratio with median in the denominator undefined. However, the current definition should assess the risk severity for your own clinical trial data. As this time, it is not possible to incorporate these new variables into any of the overall risk indicators. However, enterprising individuals can use JSL scripting to create new variables that are weighted averages of two or more risk indicators.

Table 3.2 Risk Thresholds for New Variables

Variable	Green	Yellow	Red
Percent Discontinued due to Death	≤10% more or less than mean value of all sites	>10% and ≤20% more or less than mean value of all sites	>20% more or less than mean value of all sites
Percent Discontinued due to AE	≤10% more or less than median value of all sites	>10% and ≤20% more or less than median value of all sites	>20% more or less than median value of all sites

Variable	Green	Yellow	Red
Percent Lost to Followup	No subjects lost to followup	>0% and ≤5% of subjects lost to followup	>5% of subjects lost to followup

Figure 3.25 New Risk Indicators Defined from JSL Script

Study Site Identifier	Country	Discontinued	Percent Discontinued of ...	Discontinued due to Death	Percent Discontinued due to Death	Discontinued due to AE	Percent Discontinued due to AE	Lost to Followup	Percent Lost to Followup
1 01	USA	12	23.5	9	17.6	0	0.0	3	5.9
2 02	USA	5	15.6	3	9.4	0	0.0	2	6.3
3 03	USA	7	30.4	2	8.7	0	0.0	5	21.7
4 04	FRA	6	23.1	6	23.1	0	0.0	0	0.0
5 05	ITA	4	80.0	4	80.0	0	0.0	0	0.0
6 06	USA	3	42.9	3	42.9	0	0.0	0	0.0
7 07	CHN	2	40.0	2	40.0	0	0.0	0	0.0
8 08	GBR	9	39.1	9	39.1	0	0.0	0	0.0
9 09	CAN	13	32.5	13	32.5	0	0.0	0	0.0
10 10	USA	5	29.4	5	29.4	0	0.0	0	0.0
11 12	DEU	7	43.8	7	43.8	0	0.0	0	0.0
12 14	CAN	7	9.3	7	9.3	0	0.0	0	0.0
13 16	USA	1	2.6	1	2.6	0	0.0	0	0.0
14 17	USA	5	27.8	5	27.8	0	0.0	0	0.0
15 18	JPN	1	4.5	1	4.5	0	0.0	0	0.0
16 19	CAN	3	33.3	2	22.2	0	0.0	1	11.1
17 20	FRA	3	16.7	3	16.7	0	0.0	0	0.0
18 21	ESP	2	20.0	2	20.0	0	0.0	0	0.0
19 22	CAN	5	21.7	5	21.7	0	0.0	0	0.0
20 23	USA	7	24.1	7	24.1	0	0.0	0	0.0
21 24	CHN	2	9.5	2	9.5	0	0.0	0	0.0
22 25	USA	6	40.0	6	40.0	0	0.0	0	0.0
23 26	DEU	1	12.5	1	12.5	0	0.0	0	0.0
24 27	CHE	5	20.8	3	12.5	0	0.0	2	8.3
25 28	CAN	9	12.2	9	12.2	0	0.0	0	0.0
26 29	USA	4	33.3	3	25.0	0	0.0	1	8.3
27 30	USA	0	0.0	0	0.0	0	0.0	0	0.0
28 31	ITA	3	60.0	3	60.0	0	0.0	0	0.0
29 32	USA	12	26.7	9	20.0	0	0.0	3	6.7
30 33	USA	2	22.2	2	22.2	0	0.0	0	0.0
31 34	CHN	2	28.6	2	28.6	0	0.0	0	0.0
32 35	USA	1	12.5	1	12.5	0	0.0	0	0.0
33 36	USA	0	0.0	0	0.0	0	0.0	0	0.0
34 37	ESP	3	23.1	2	15.4	0	0.0	1	7.7
35 39	USA	4	23.5	3	17.6	0	0.0	1	5.9
36 40	GBR	6	15.8	5	13.2	0	0.0	1	2.6
37 42	GBR	3	15.0	3	15.0	0	0.0	0	0.0
38 44	USA	8	21.1	7	18.4	0	0.0	1	2.6
39 45	JPN	3	14.3	2	9.5	0	0.0	1	4.8
40 46	USA	5	15.6	5	15.6	0	0.0	0	0.0

This script can be used to add new variables every time the RBM analysis is updated. However, the method described above is a bit clunky to use, and difficult to share with users who are less comfortable with manipulating code. To create a JMP Add-In so that these variables can easily be created by any user, go to **File > New > Add-In**. On the first tab, name the Add-In and provide some additional background (Figure 3.26 on page 112). On the second tab, Click **Add Submenu** to add a **Risk-Based Monitoring** submenu, and **Add Command** to add a command within this menu (Figure 3.27 on page 113).

Figure 3.26 JMP Add-In to Create Risk Indicators

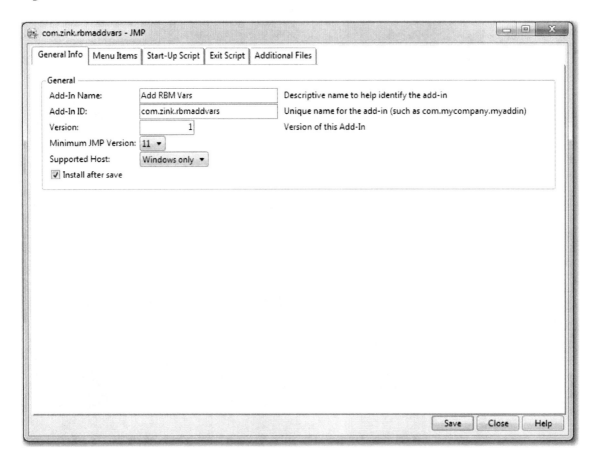

I named the command **Add RBM Variables**. For ease, I copied and pasted the script into the **Action** area. Click **Save** and choose a location to save the new Add-In (based on the **Add-In ID**, my file would be called com.zink.rbmaddvars.jmpaddin). Based on the selected option on the first tab, this menu item will be added to the **Add-In** menu once the file is saved. You can apply this command to future RBM analyses by going to **Add-In > Risk-Based Monitoring > Add RBM Variables**. Further, this file can be shared with any colleagues through email; double-clicking on the file will install this new command to their personal instance of JMP Clinical. If new variables are required, simply update the script and create a new version of the Add-In with the same name; installing the new script will overwrite the previous version of the Add-In. Users looking for further detail or assistance with JSL scripting can to **Help > Scripting Index**, or **Help > Books > Scripting Guide**, or **Help > Books > JSL Syntax Reference**. Help for Add-Ins is available in Chapter 15 of the **Scripting Guide**.

Of course, JMP Add-ins can be defined for any of the analyses or graphics suggested in Section 3.3 or from the analyst's imagination. The goal is to arrange the output from any JMP platform according to user preference with any additional options turned on. Then, click the red triangle menu and then **Script > Save Script to Script Window**. From here, you can add the line dt = Current Data Table() and then proceed to create an Add-In as described above. For example,

Figure 3.28 on page 114 contains the script that creates the bar chart in Figure 3.22 on page 105. In the next section, we'll define an Add-In to calculate additional risk indicators for adverse events and generate summary figures of these new variables.

Figure 3.27 Risk-Based Monitoring Submenu and Add RBM Vars Command

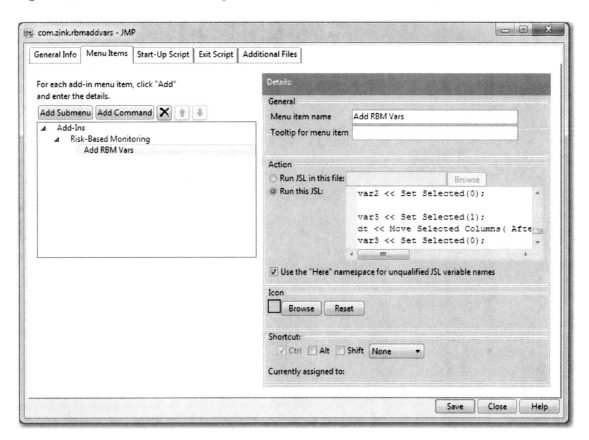

Figure 3.28 *JSL Script for Bar Chart for Overall Risk Indicator by Study Site Identifier Grouped by Monitor*

```
dt = Current Data Table();

Graph Builder(
    Size( 1740, 918 ),
    Show Control Panel( 0 ),
    Variables(
        X( :Monitor ),
        X( :Study Site Identifier, Position( 1 ) ),
        Y( :Overall Risk Indicator )
    ),
    Elements(
        Bar(
            X( 1 ),
            X( 2 ),
            Y,
            Legend( 2 ),
            Bar Style( "Side by side" ),
            Summary Statistic( "Mean" )
        )
    )
);
```

3.4.2 New Adverse Event Variables and Figures

TransCelerate BioPharma recently released an update to their position paper on RBM [3]. In Section 2.3.2, a table summarizes pilot metrics that will be used to implement and test suggested RBM methodologies. Among the listed risk indicators is the proportion of unreported but confirmed SAEs. Whether an SAE is reported or not may require information that exists outside the study database. However, it is possible to make use of data captured in the study database to assess whether or not sites have an overabundance or lack of events meeting the criteria for seriousness (e.g. death, hospitalization, disability) [4]. For example, a scarcity of SAEs compared to the overall frequency of adverse events could suggest sites that may not be reporting the seriousness of events appropriately. As part of this exercise, we'll also derive the frequency of nonserious AEs by subtracting serious events from the total number of events experienced. Below is a script (Example Code 3.2 on page 115) which can be used to calculate the **Total Nonserious AEs on Study**, the **Average Nonserious AEs per Randomized Subject**, and **Percent SAEs**, which is the proportion of AEs that meet the criteria for seriousness (Figure 3.29 on page 116, uses **Cols > Reorder Columns**). While I don't apply risk thresholds to identify mild, moderate, or severe risk in this particular example, users are free to do so using what they've learned from "3.4.1 New Discontinuation Variables and Risk Thresholds" on page 108. Using **Cols > Reorder Columns > Move Selected Columns** to move these new columns among the adverse event variables naturally includes them in the **Adverse Events** group. Alternatively, adding the line dt << Group Columns("Adverse Events", {Total Nonserious AEs on Study, Average Nonserious AEs per Randomized Subject,Percent SAEs}) will add the new variables to the **Adverse Events** group.

Example Code 3.2 Script to Add Adverse Event Variables

```
Names Default to Here(1);
dt = Current Data Table();

// New Variables
name1 = "Total Nonserious AEs on Study";
name2 = "Average Nonserious AEs per Randomized Subject";
name3 = "Percent SAEs";
dt << New Column( name1, Formula( :Total AEs on Study - :Total SAEs on Study));
dt << New Column( name2, , Format( "Fixed Dec", 8, 2),
        Formula( :Total Nonserious AEs on Study/:Randomized));
dt << New Column( name3, , Format( "Fixed Dec", 8, 1),
        Formula( :Total SAEs on Study/:Total AEs on Study * 100));
var1 = Column(dt,name1);
var2 = Column(dt,name2);
var3 = Column(dt,name2);
```

Figure 3.29 New Adverse Event Variables

	Study Site Identifier	Country	Total Nonserious AEs on Study	Average Nonserious AEs per Randomized Subject	Percent SAEs
1	01	USA	483	9.47	9.6
2	02	USA	105	3.28	11.0
3	03	USA	137	5.96	6.8
4	04	FRA	150	5.77	14.3
5	05	ITA	35	7.00	7.9
6	06	USA	76	10.86	16.5
7	07	CHN	67	13.40	5.6
8	08	GBR	262	11.39	12.1
9	09	CAN	182	4.55	10.8
10	10	USA	82	4.82	19.6
11	12	DEU	139	8.69	10.9
12	14	CAN	249	3.32	10.4
13	16	USA	166	4.26	2.9
14	17	USA	223	12.39	10.1
15	18	JPN	78	3.55	9.3
16	19	CAN	30	3.33	14.3
17	20	FRA	130	7.22	8.5
18	21	ESP	91	9.10	12.5
19	22	CAN	108	4.70	33.7
20	23	USA	183	6.31	8.5
21	24	CHN	111	5.29	8.3
22	25	USA	157	10.47	10.8
23	26	DEU	28	3.50	6.7
24	27	CHE	112	4.67	15.2
25	28	CAN	523	7.07	11.5
26	29	USA	109	9.08	13.5
27	30	USA	13	2.60	7.1
28	31	ITA	23	4.60	17.9
29	32	USA	322	7.16	11.5
30	33	USA	49	5.44	10.9
31	34	CHN	50	7.14	24.2
32	35	USA	64	8.00	7.2
33	36	USA	86	14.33	1.1
34	37	ESP	88	6.77	3.3
35	39	USA	85	5.00	15.8
36	40	GBR	240	6.32	6.3
37	42	GBR	123	6.15	9.6
38	44	USA	276	7.26	6.8
39	45	JPN	77	3.67	1.3
40	46	USA	276	8.63	5.8

Run the **Risk-Based Monitoring** report for **Nicardipine** using the **Default Risk Threshold** data set, and click anywhere within the **Site-Level Risk Indicator** data table to make it the current data table. Run the **Script for AEs.jsl** (Figure 3.30 on page 118, available from the companion website; drag onto JMP Clinical starter to open). Let's generate some summary figures using these new risk indicators. Go to **Graph > Graph Builder**. Drag **Study Site Identifier** to **X** and **Percent SAEs** to **Y**; click on the bar icon to create a bar chart. Click **Done** to close the control panel. Open a second instance of **Graph Builder** by going to **Graph > Graph Builder**. Drag **Study Site Identifier** to **X** and **Average AEs per Randomized Subject** to **Y**. Right click in the figure and click **Add > Line** to connect the points in the graph. Next, drag **Average SAEs per Randomized Subject** to **Y** towards the lower corner to get a separate line plot for SAEs; do the same for **Average Nonserious AEs per Randomized Subject** (Figure). Change the title by clicking on it and entering the text **Adverse Event Averages per Randomized Subject vs. Study Site Identifier**. For both figures, click the red triangle menu and then **Script > Save Script to Script Window**. Copy the JSL code and paste it into the script in which the new AE variables were created. You should obtain something similar to Example Code 3.3 on page 118, though I added the dt << prior to the Graph Builder calls to be a bit more explicit about which data I am using for the platform (**Script for AEs with Figures.jsl** is available from the companion site). This script can be applied to both the site- and country-level data tables. Interested users can create a JMP Add-In using this script as described in the previous section.

Figure 3.30 *Building a Graph for Adverse Events*

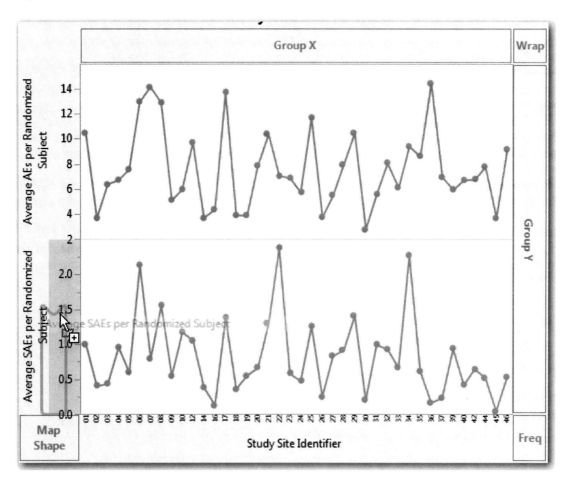

Example Code 3.3 *Script for Generating AE Variables and Figures*

```
Names Default to Here(1);
dt = Current Data Table();
// New Variables
name1 = "Total Nonserious AEs on Study";
name2 = "Average Nonserious AEs per Randomized Subject";
name3 = "Percent SAEs";
dt << New Column( name1, Formula( :Total AEs on Study - :Total SAEs on Study));
dt << New Column( name2, , Format( "Fixed Dec", 8, 2),
       Formula( :Total Nonserious AEs on Study/:Randomized));
dt << New Column( name3, , Format( "Fixed Dec", 8, 1),
       Formula( :Total SAEs on Study/:Total AEs on Study * 100));
var1 = Column(dt,name1);
var2 = Column(dt,name2);
var3 = Column(dt,name2);
```

```
// Graph Builder Line Plot
dt << Graph Builder(
    Show Control Panel( 0 ),
    Variables(
        X( :Study Site Identifier ),
        Y( :Average AEs per Randomized Subject ),
        Y( :Average SAEs per Randomized Subject ),
        Y( :Average Nonserious AEs per Randomized Subject )
    ),
    Elements(Position( 1, 1 ), Points( X, Y, Legend( 3 ), Jitter( 1 ) ),
        Line( X, Y, Legend( 4 ), Row order( 0 ), Summary Statistic( "Mean" ) )),
    Elements(Position( 1, 2 ), Points( X, Y, Legend( 1 ), Jitter( 1 ) ),
        Line( X, Y, Legend( 5 ), Row order( 0 ), Summary Statistic( "Mean" ) )),
    Elements(Position( 1, 3 ), Points( X, Y, Legend( 7 ), Jitter( 1 ) ),
        Line( X, Y, Legend( 8 ), Row order( 0 ), Summary Statistic( "Mean" ) )),
    SendToReport(Dispatch({},"graph title",TextEditBox,
        {Set Text("Adverse Event Averages per Randomized Subject
        vs. Study Site Identifier")})
    )
);

// Graph Builder Bar Chart
dt << Graph Builder(
    Show Control Panel( 0 ),
    Variables( X( :Study Site Identifier ), Y( :Percent SAEs ) ),
    Elements(Bar(X,Y,Legend( 3 ), Bar Style( "Side by side" ),
        Summary Statistic( "Mean" )))
);
```

While it is useful to generate multiple plots from a single Add-In, managing multiple figure windows can be a little tedious. The **Script for AEs with Single Figure Window.jsl** (downloadable from companion site) illustrates how to combine output horizontally from calls to multiple platforms using **New Window** and H List Box() (Figure 3.31 on page 120). Alternatively, V List Box() can be used to stack figures vertically. Any number of figures can be combined in this way. Just be sure to separate each platform call with a comma! From the figure, we can see that almost 35% of SAEs from Site 22 are considered serious, while Sites 36 and 45 have almost no SAEs to speak of. These sites may benefit from further scrutiny.

Figure 3.31   Adverse Event Figure Window

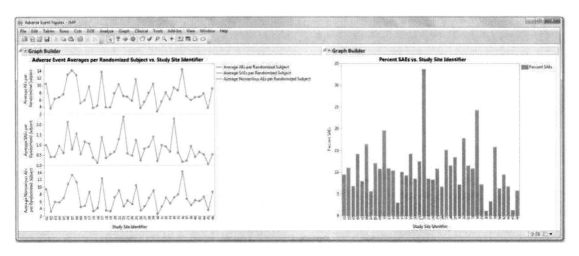

3.4.3 Analyses at the Monitor Level

In this last example, we'll examine summary analyses conducted at the monitor level. By default, the analyses conducted within **Risk-Based Monitoring** summarize at the site and country level. Summarizing at the country level provides the analyst insight into whether certain differences in medical practice or the regulatory landscape between countries impact the safety and quality measures under investigation. Using similar logic, one could conclude that a trial monitor's experience, thoroughness, and willingness to follow-up with clinical sites may play a role in the quantity and severity of safety and quality findings observed. While platforms such as **Graph Builder** have the capacity to sum or average findings across sites within monitor, I'll illustrate a new tool to build a data set at the monitor level to assess whether or not any important differences exist between monitors. Similar to the previous section, I won't spend time applying risk thresholds to the variables; users interested in doing so can refer to the **For Each Row** function in Example Code 3.1 on page 108.

Figure 3.32 Tabulate Platform

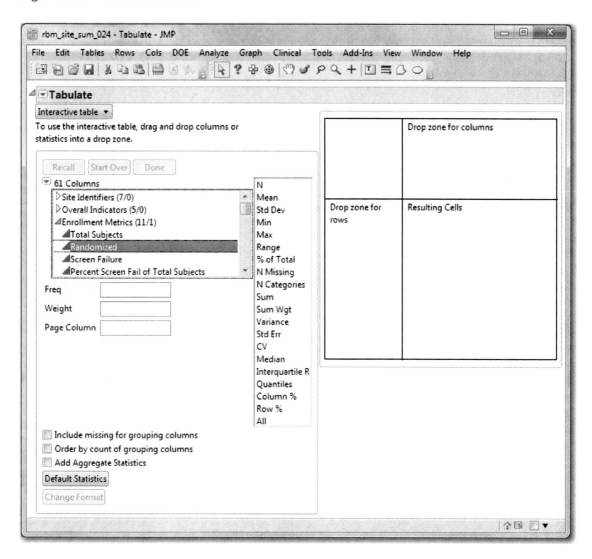

Figure 3.33 Overall Sums of Manually Entered Risk Indicators

	Randomized	Total Queries	Overdue Queries	Missing Pages	Computed Deviations	Computed Eligibility Violations
	Sum	Sum	Sum	Sum	Sum	Sum
	906	7565	485	29	163	589

Run the **Risk-Based Monitoring** report for **Nicardipine** using the **Default Risk Threshold** data set, and click anywhere within the **Site-Level Risk Indicator** data table to make it the current data table. Go to **Analyze > Tabulate**. Like **Graph Builder**, the **Tabulate** platform provides a drag and drop environment to summarize data (Figure 3.32 on page 121). Drag **Randomized**,

Total and **Overdue Queries**, **Missing Pages** and **Computed Deviations** and **Eligibility Violations** to the column drop zone (Figure 3.33 on page 121). This provides the sum of these variables across all 40 clinical sites. To get the average of **CRF Entry Response Time** and **Query Response Time**, first drag the summary statistic **Mean** just past the last column to create a **Mean** column (Figure 3.34 on page 122).

Figure 3.34 Adding Columns for Means

Randomized	Total Queries	Overdue Queries	Missing Pages	Computed Deviations	Computed Eligibility Violations
Sum	Sum	Sum	Sum	Sum	Sum
906	7565	485	29	163	589

Now drag **CRF Entry Response Time** and **Query Response Time** to this mean column to compute the averages across the 40 sites. Finally, drag **Monitor** to the first blank column to produce sums and averages within each monitor (Figure 3.35 on page 122). Click **Done** to finish the table. Go to the red triangle, then **Script > Save Script to Script Window**. This script can be used to create a monitor-level data table Add-In (Example Code 3.4 on page 122). **Script for Monitors.jsl** is available from the companion site. In this script, I use the **Tabulate** call we just created with the addition of <<Make Into Data Table(Output Table("Monitor Summary") to output the data into a data table named **Monitor Summary**. The table << Close Window closes the Tabulate output window once the data table is created. The line dtm = Data Table("Monitor Summary") creates a reference to more easily refer to our new data table. Subsequent lines should look familiar from previous examples: We create two risk indicators for queries and use **Distribution** to summarize them. Notice here the use of the **Name** function when referring to some of the variables. This is necessary to mask the parentheses so they are not interpreted as part of a formula. Any number of analyses are possible; users are encouraged to modify the **Script for Monitors.jsl** to add summary figures prior to creating their own Add-In (Figure 3.36 on page 123).

Figure 3.35 Manually Entered Risk Indicators at the Monitor Level

Monitor	Randomized Sum	Total Queries Sum	Overdue Queries Sum	Missing Pages Sum	Computed Deviations Sum	Computed Eligibility Violations Sum	Query Response Time Mean	CRF Entry Response Time Mean
Monitor A	162	955	42	1	13	58	4.390	5.782
Monitor B	136	1325	38	1	13	40	5.674	9.780
Monitor C	69	417	26	3	16	46	4.419	1.529
Monitor D	34	221	59	3	8	44	3.105	4.971
Monitor E	57	504	33	1	12	47	4.645	2.416
Monitor F	79	412	32	3	17	41	10.289	2.894
Monitor G	87	1103	55	6	26	61	6.588	0.937
Monitor H	34	247	37	3	8	54	5.913	8.837
Monitor I	91	702	56	2	16	77	8.306	4.922
Monitor J	81	972	30	3	14	45	5.656	8.144
Monitor K	33	256	45	2	10	37	6.382	6.294
Monitor L	43	451	32	1	10	39	4.297	4.954

Example Code 3.4 Script to Create Data Table and Analysis at the Monitor Level

```
Names Default To Here( 1 );
dt = Current Data Table();
```

```
// Tabulate to calculate summaries at Monitor Level
table = dt << Tabulate(  // setting reference for tabulate platform
    Add Table(
        Column Table(
            Analysis Columns(:Randomized,:Total Queries,:Overdue Queries,:Missing Pages,
                :Computed Deviations,:Computed Eligibility Violations)
        ),
        Column Table(
            Analysis Columns( :Query Response Time, :CRF Entry Response Time ),
                Statistics( Mean )
        ),
        Row Table( Grouping Columns( :Monitor ) )
    ), // Note the comma. If missing will cause an error
    <<Make Into Data Table( Output Table( "Monitor Summary" ) )
        // Added to save tabular output to data table
);

table << Close Window; // Close tabulate summary
dtm = Data Table( "Monitor Summary" ); // Define reference for monitor data table

// New Variables
name1 = "Average Queries per Randomized Subject";
name2 = "Average Overdue Queries per Randomized Subject";
// Note use of Name() function to mask parentheses in variable name
dtm << New Column( name1, Format( "Fixed Dec", 8, 2 ),
        Formula(:Name( "Sum(Total Queries)" )/:Name( "Sum(Randomized)" )));
dtm << New Column( name2,Format( "Fixed Dec", 8, 2 ),
        Formula(:Name( "Sum(Overdue Queries)" )/:Name( "Sum(Randomized)" )));

// Distribution
dtm << Distribution(
    Continuous Distribution( Column( :Average Queries per Randomized Subject ) ),
    Continuous Distribution( Column( :Average Overdue Queries per Randomized Subject ) )
);
```

Figure 3.36 New Add-Ins for Risk-Based Monitoring

3.5 Final Thoughts

In this chapter, I hope to have inspired the imagination of analysts and reviewers to the wide variety of additional functionality available within the **Analyze**, **Graph**, and **Clinical > Pattern Discovery** menus. One topic I didn't address in great detail is the ability to generate models using the **Fit Model** platform, or to use tools contained within the **Modeling** and **Clinical > Predictive Models** menus. For example, I could fit Poisson models using **Fit Model**, including a log (**PatientWeek**) offset to model the frequency of adverse events while accounting for varying exposures using risk indicators captured at the subject level. Or I could use **Modeling > Partition** to produce binary trees to uniquely classify risk severity categories for sites or countries based on the available covariates. Finally, the **Clinical > Predictive Models > Predictive Modeling Review** report can be used to define predictive modeling "experiments" to compare across multiple predictive models and tuning parameters, reduce the covariate space by selecting representative variables from covariate clusters, or apply cross-validation techniques to assess model performance. The reports within **Clinical > Predictive Models > Predictive Modeling Review** are specifically designed for the wide (more variables than observations) data scenarios typically seen in the life sciences.

Speaking of models, I expect this to be an area of future development for the RBM functionality of JMP Clinical. While it is certainly useful to identify risk as it occurs, it would be more effective to predict when risk thresholds will be crossed to intervene before safety or quality becomes a concern (similar to what is suggested in [5]). Further development of JMP Clinical will include functionality to summarize risk over time to provide better insight into how risk changes over the course of the clinical trial. In addition, there will be some work performed to incorporate results from the fraud detection analyses described in Chapters 4 and 5 into the RBM analysis and review. Finally, we anticipate that users would like to schedule regular RBM analysis and review to coincide with updates to their clinical database. We envision functionality to schedule reviews at user-defined intervals. Of course, user feedback will determine the features to be implemented and their priority.

References

1. TransCelerate BioPharma Inc. (2013). Position paper: Risk-based monitoring methodology. Available at: http://transceleratebiopharmainc.com/.

2. Mahalanobis PC (1936). On the generalised distance in statistics. *Proceedings of the National Institute of Sciences of India* 2: 49–55.

3   TransCelerate BioPharma Inc. (2014). Risk-based monitoring update – volume I. Available at: http://www.transceleratebiopharmainc.com/wp-content/uploads/2014/01/TransCelerate-RBM-Update-Volume-I-FINAL-27JAN2014.pdf

4   International Conference of Harmonisation. (1995). E3: Structure and Content of Clinical Study Reports. Available at: http://www.ich.org/fileadmin/Public_Web_Site/ICH_Products/Guidelines/Efficacy/E3/E3_Guideline.pdf.

5   Pogue JM, Devereaux PJ, Thorlund K & Yusuf S. (2013). Central statistical monitoring: Detecting fraud in clinical trials. *Clinical Trials* 10: 225-235.

4
Detecting Fraud at the Clinical Site

4.1 Introduction	**127**
4.2 Study Visits	**129**
4.2.1 Weekdays and Holidays	129
4.2.2 Study Scheduling	136
4.3 Measurements Collected at the Clinical Site	**144**
4.3.1 Tests with No Variability	144
4.3.2 Duplicate Sets of Measurements	148
4.3.3 Digit Preference	154
4.3.4 A Brief Interlude from the Fraud Detection Menu	157
4.4 Multivariate Analyses	**164**
4.4.1 Multivariate Inliers and Outliers	164
4.4.2 Hierarchical Clustering of Subjects Within Clinical Sites	168
4.5 Final Thoughts	**175**
References	**176**

4.1 Introduction

Fraud in clinical trials is thought to be rare, though its prevalence is likely underestimated due to previously unavailable or limited tools and training for diagnosis, and fear over negative publicity [1]. However, several examples of fraudulent activity are available in the literature. Perceived misconduct was identified in a trial examining the effectiveness of a dietary intervention on reducing the risk of cardiovascular deaths [2]. After failing to get cooperation from the manuscript author to address perceived inconsistencies, a detailed analysis of the data was performed that identified major differences in baseline characteristics and digit preference among the trial arms

[3]. In the National Surgical Adjuvant Breast and Bowel Project, approximately 100 patients from a clinical site in Montreal were discovered to be ineligible; the patients' data had been manipulated so that they appeared to meet entry criteria [1,4]. A review of findings from an oncology trial was conducted in South Africa by investigators who were provided with data for only three-quarters of the high-dose chemotherapy patients. Based on available information, reviewers identified improper documentation of eligibility criteria and no evidence of informed consent [4]. In a large cardiovascular trial, 9 out of 196 sites representing nearly 1,000 patients were identified to have fabricated various aspects of trial data; the authors cite further examples of fraud from several other cardiovascular studies [5].

Other examples of fraud from scientific research include data manipulation from a case study lacking suitable controls; it linked the measles, mumps, and rubella vaccine to autism [6]. Data falsification was identified at a laboratory in a multicenter animal study sponsored by the National Heart, Lung, and Blood Institute [7]. Research of transplantation immunology conducted using white and black mice was performed by a researcher whose best examples of skin transplants from black to white mice were the result of a permanent ink marker [8]. Genomic data from oncology research was deemed irreproducible, perhaps discovered only because the data and methods were available in the public domain [9-11]. Two recent publications describe higher-than-expected rates of scholarly retractions in the life science and biomedical literature, often due to fraud or suspected fraud [12,13].

In addition to the literature, we may have first-hand knowledge of trial misconduct, or have heard accounts from our colleagues. For example, a fellow statistician once recalled for me a clinical trial in which the data retrieved from a particular data collector at a site was fabricated. Multiple measurements were required at each visit, and the collector completed each set of data with values randomly generated from a single actual measurement collected at each visit.

Fortunately, trial integrity will be preserved even in the presence of incorrect or manipulated data, most often due to randomization and blinding of study medication, or because the anomalies are limited to a small number of sites [1,4,14-16]. Further, identifying and documenting fraud can be a lengthy and expensive process that, once initiated, can damage the perception and reputation of a research institution [2,4,6,11]. So as in Chapter 1, I ask, why bother looking for fraud at all? The unfortunate death of Julie Jacobs serves as a reminder: to protect the well-being of study participants [10]. If analysis of data quality, including methods to detect potentially fraudulent data, were applied early and often enough, problems could be identified and addressed early, likely limiting any serious outcomes. And while this is speculation, perhaps this proactive effort coupled with limited wrongdoing would limit the negative perception of a sponsor should the identified misconduct be made public.

Chapters 4 and 5 summarize various graphical and statistical approaches used to identify site- or patient-perpetrated fraud, respectively. As we walk through the examples, you will appreciate that in many instances, there are no black-or-white, yes-or-no answers as to whether a particular finding is due to malicious intent, carelessness, or some other quality issue – or whether the unusual result is valid with a perfectly reasonable explanation! Background in a particular therapeutic area or knowledge of particular study tests or procedures is useful to explain any data anomalies, but as Evans points out, the answer to whether a finding is truly fraudulent or not likely lies outside of the study database, often requiring further communication and investigation with the study site [17]. However, detecting unusual patterns in the data is a useful means to guide study monitors in their daily activities so that their time is better spent [14]. Below, I describe analyses that examine various facets of study scheduling ("4.2 Study Visits" on page 129), analyses of findings that are collected at the study site ("4.3 Measurements Collected at the Clinical Site" on page 144), and analyses that can potentially take advantage of the totality of available data for study subjects, including their adverse experiences, medical history, and concomitant medications ("4.4 Multivariate Analyses" on page 164). In general, clinical trial data are highly structured and this structure is known when using CDISC. Further, human beings are bad at fabricating realistic data; we'll exploit these facts in the pages that follow [1,17,18]. The majority of the analyses I describe below are located in the **Fraud Detection** menu of the Clinical Starter. Please note: unlike the example in Chapters 2 and 3 for risk-based monitoring that shows a global reach for sites in the **Nicardipine** trial, the remainder of the book shows sites belonging to the United States (the original and correct country specification).

4.2 Study Visits

4.2.1 Weekdays and Holidays

Sometimes we are so wrapped up in the day-to-day activities of conducting the clinical trial that we forget to perform some of the more obvious data checks. Data-cleaning activities that seem unlikely to bear fruit because "I can't believe they would do this" or "there is no way they could mess this up" are always worth performing, especially when they are straightforward to implement. When someone fabricates or manipulates data, there are many aspects to consider so that the misdeed goes undetected; one detail may go overlooked or other facets may not even be considered! Consider, for example, the visit dates themselves. Sure, we may have validation checks to ensure that Visit (k-1) occurs before Visit k for all values of k, but how often do we bother examining the significance of the actual day?

Clinical trials take place in the real world and are subject to the forces that govern them, even those that are seemingly less important than the disease under investigation. Consider weekends and holidays, and ask yourself the following questions. Is it likely that a trial participant would show up for a study visit on a major holiday? Would a physician's office be open on a Saturday or Sunday? Now consider natural disasters. How likely is it for a clinical site to have conducted study visits immediately after a hurricane caused major flooding, disrupting supply lines and travel? How about a blizzard that generated a foot of snow in an area unequipped for such rare occurrences? What about past public health events over swine or avian flu, or other outbreaks that may have stretched healthcare resources to the limit? Is it likely for scheduling at study sites located in these areas to go unaffected?

The **Weekdays and Holidays** report examines the dates on which demographic characteristics and medical history are collected (CDISC variables DM.DMDTC and MH.MHDTC, respectively), as well as the dates over which study visits occur (dates between and including SV.SVSTDTC and SV.SVENDTC). If the Study Visits (SV) domain is unavailable, all CDISC findings domains are collected to identify the set of dates where measurements were either recorded or specimens collected (xx.xxDTC, where xx refers to a specific CDISC domain). In general, the analysis will identify the day of the week on which the visit occurs and whether the visit coincides with a major

U.S. or Canadian holiday, though some holidays such as Christmas or New Year's Day are more universal and are applied to every site. Holidays are flagged depending on the country in which the clinical site is located. For example, the U.S. holiday of Thanksgiving (the fourth Thursday in November) will be flagged only for sites within the U.S. While initially this may seem limiting, users can define custom holidays and events (e.g., severe weather) for a specific country or set of countries using **Add Holiday or Event**. I describe this particular functionality a bit later. For now, we will review an example for **Nicardipine**; click **Weekdays and Holidays > Run**. There are no options for this report save for applying filters to subset the analysis to a specific study population or set of subjects.

Results are presented in Figure 4.1 on page 131 using histograms taken from the **Distribution** platform. The following variables are summarized: Holiday, Weekday, Country, Study Site Identifier, and Date Label (the date demography or medical history is collected, or visit date). If visits are determined to occur over multiple days from the SV domain, visit days are distinguished between starting, ending, and days that occur between using the Date Label; otherwise, single-day visits are listed as Visit Start Dates. Drill downs available for this analysis allow users to **Show Subjects** or **Profile Subjects** with unusual visits, **Create Subject Filter** to subset subsequent analyses to the subjects with unusual visits, or **Cluster Subjects** based on data available from other domains. A **Data Filter** allows the analyst to subset the histogram by numerous criteria; other variables may be added to the filter by clicking **AND** or **OR**. Notice how the histogram for Weekday shows a rather uniform distribution over the days of the week. Considering these subjects experienced a subarachnoid hemorrhage (SAH), it is not unusual for visits to occur on the weekend. However, for trials in some therapeutic areas it may be extremely unlikely for visits to occur over the weekend. In these instances, if visits do occur on weekends, this may be infrequent or very rare. But how can we quickly review the results for each site? One possibility is to generate a contingency table using **Analyze > Fit Y by X**. Alternatively, we can take advantage of the **Animation** feature under the red triangle menu of the **Data Filter**. First, let's click the red triangle for Weekday and then **Order By > Default** so that the days of the week are not sorted by frequency when the histograms are subset for each site in turn. Then, turn on the **Animation** feature to open the **Animation Controls** in the **Data Filter** (Figure 4.2 on page 132). Click the outline box around Study Site Identifier in the filter so that the animation is applied to clinical sites (it should turn blue when selected, as in the figure). It may be worthwhile to also add Date Label to the **Data Filter** using **AND** so that you can subset dates specifically to those on which a visit occurred.

Figure 4.1 Distributions of Weekdays and Holidays

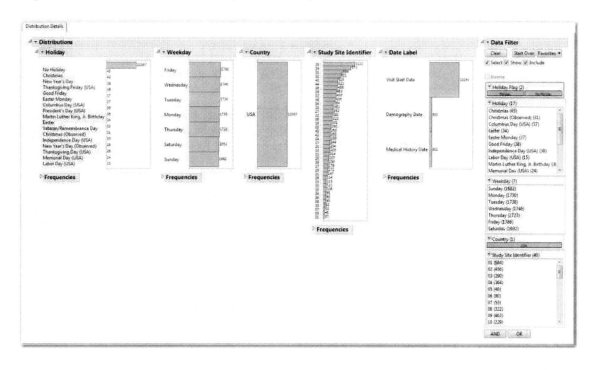

Figure 4.2   Data Filter With Study Site Identifier Highlighted

Clicking the "play" button will cycle through the list of sites one after the other, subsetting the results of the histograms to the selected center. Now, you just need to watch for any unusual pattern. In general, there doesn't appear to be any study center with an overabundance of Saturdays or Sundays. However, let's perform a test to see if the distribution of Visit Days varies according to Study Center. Click **Clear** in the **Data Filter** to remove any subsetting, and then subset records to Visit Start Date in the **Data Filter**. Now go to **Tables > Subset > OK**. This action will subset the invisible data table providing results for the **Weekdays and Holidays** Report to those records that correspond to a study visit (Visit Start Date). Now go to **Analyze > Fit Y by X** and set Weekday as **Y, Response** and Study Site Identifier as **X, Factor**; click **OK**. A mosaic plot, contingency table, and chi-square test are provided (Figure 4.3 on page 133).

Based on the p-value of 1, there is clearly no relationship between clinical site and weekday, though this is rather obvious from the graphical depiction in the mosaic plot. Perform a similar exercise to assess whether the distribution of Holiday Flag differs across the clinical sites for study visits. You'll notice a significant p-value. However, if you assess the expected cell counts (through the red triangle), you'll notice that many values are less than 5, calling into question the validity of the asymptotic statistical test. In general, if in doubt about the appropriateness of this or any test, consult a statistician. As a note, mosaic plots are available from **Graph > Graph Builder**.

Figure 4.3  Mosaic Plot of Weekday by Study Site Identifier

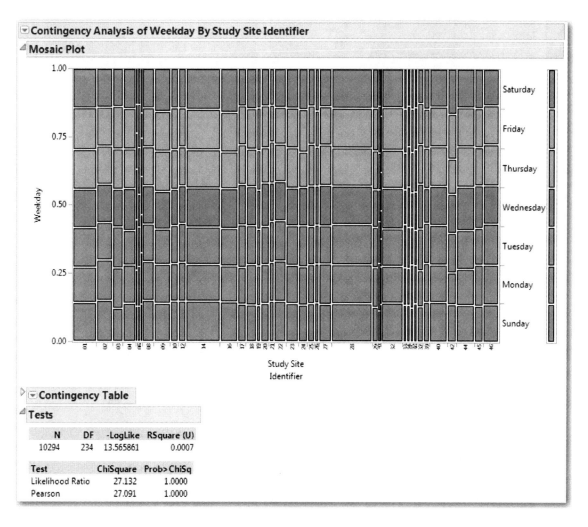

Now that I have illustrated some of the default functionality of the **Weekdays and Holidays** Report, let's make use of **Add Holiday or Event** to identify clinical visits that occurred during severe weather events. Visits for the **Nicardipine** trial took place between October 1987 and September 1989. I'll define two severe weather events based upon major hurricanes that hit the United States. The first was hurricane Gilbert; it traveled across the Gulf of Mexico and hit the

134 Chapter 4 / Detecting Fraud at the Clinical Site

Midwest states from September 16 to 19, 1988. The second hurricane, Hugo, hit the South Carolina coast on September 21, 1989, and traveled north toward Lake Erie, where it dissipated on the 22nd. Clicking **Add Holiday or Event** will open the dialog shown in Figure 4.4 on page 134. Users have the ability to name the event, choose a start and stop date, and select the countries affected by the event or where the holiday is celebrated. While it is possible to type the date into the cells directly (two-digit day, abbreviated month spelled out, four-digit year), hovering over the left side of the cell with the cursor will display a blue triangle that, when pressed, will open a calendar widget to more easily enter the date (Figure 4.5 on page 135). Any holidays or events defined will be applied to all studies where visits occur on the appropriate dates from the selected countries.

Figure 4.4 Create Holiday or Event Dialog

Figure 4.5 JMP Calendar Widget

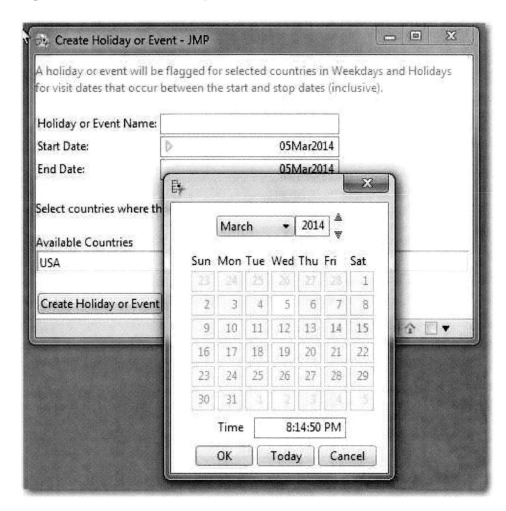

Figure 4.4 on page 134 displays an entry for Hurricane Gilbert, and I have selected USA as the country affected by the hurricane since this is where my clinical sites are located. Countries available for selection will be based upon the studies registered within JMP Clinical. As studies are added or data updated, a cumulative list of countries is saved to a database for ease of selection. Below, I describe how it is possible to add events for countries that have yet to be added to this database.

Once all data are supplied, click **Create Holiday or Event** to save the event to a database. Perform the same exercise to add an event for Hurricane Hugo. Running **Weekdays and Holidays** and subsetting to Holidays and Visit Start Dates using the **Data Filter** will provide the list of holidays displayed in Figure 4.6 on page 136. Notice that 60 clinic visits occurred during the time Hurricane Gilbert was over the Midwestern United States. Further investigation into the sites in which these visits occurred (from the Study Site Identifier histogram) coupled with the knowledge of the locations of these sites will signify if anything unusual has taken place. The absence of Hurricane Hugo from the list implies that no visits took place while the hurricane was inland. If data for a particular event requires updating, such as to correct a start or end date, remove countries from an event, or add countries that were not available from the drop-down dialog, users can go to **Edit Holiday or Event Data Set** to edit or add the appropriate records to the data table.

Users operating in local mode will maintain their own individual database of holidays; operating in server mode will share the set of holidays and events across all analysts. In the next section, we'll compare the distribution of the day on which each visit begins for each site compared to all others to identify any unusual patterns in visit scheduling.

Figure 4.6 Holiday Distribution Including Hurricane Gilbert

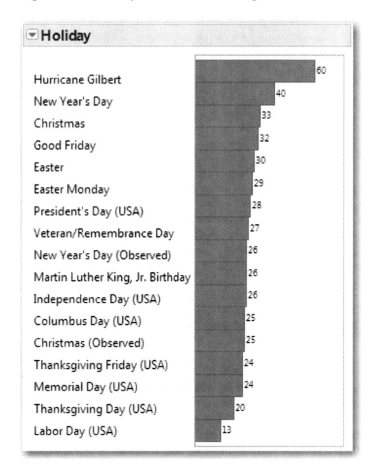

4.2.2 Study Scheduling

Other patterns that may be a cause for concern for visit dates include perfect or near-perfect visit attendance on the planned or expected study day. Figure 4.7 on page 137 presents a variation of Figure 2 in reference [4] using data provided from Buyse and coauthors [1]. For the visit summarized, the distribution for the suspect center appears to be quite different from the distribution of all other centers combined. While this may not be too unusual for sites with few subjects, the presence of the small percentages for study days 22 and 23 implies that there are a number of subjects enrolled at the suspect site. While keeping to the study schedule is important for understanding the safety and efficacy at any given point in time, such near-perfection in a world where patients forget visits or reschedule visits due to other competing priorities seems unlikely. Such an occurrence would warrant further investigation with the site in question. However, such near-perfection is only one such difference that graphs such as Figure 4.7 on page 137 can highlight. Important differences can also be due to a site being extremely off-schedule. In these cases, intervention may be necessary to identify and correct the factors contributing to the poor performance at the site so that the data collected is representative of the expected time points.

Figure 4.7 Distribution of Study Days on Which Clinic Visit Occurred

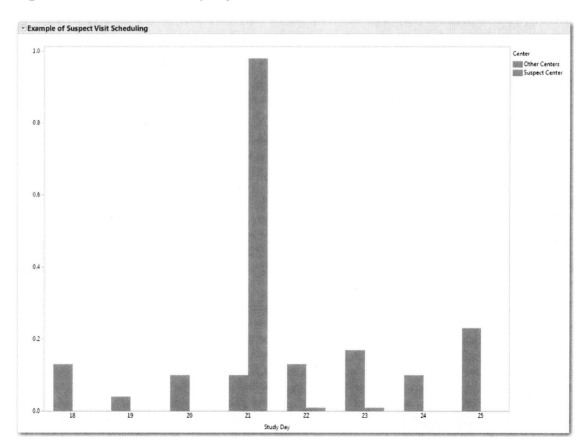

While a bar chart is useful for comparing the distribution of study day for a given visit for a site compared to all other sites, in general, a clinical trial will have many clinical sites and visits to examine for any unusual findings. In these instances, it becomes necessary to screen the database to identify unusual site-visit combinations, especially when there is no prior indication that the behavior or performance of investigators at a clinical site is in question. We can easily screen visit schedules by treating each center as the "suspect site" and comparing the distribution of each visit to that of all other sites combined. If there are i sites and j visits, this will result in up to $i \times j$ comparisons (allowing for the possibility that not all visits will occur at every site). The analysis examines the Study Day of Start of Visit (SV.SVSTDY) for each Visit (SV.VISITNUM) treating each site (DM.SITEID) as the suspect site. If SV.SVSTDY is not available, study day is computed from the Start Date/Time of Visit and the Subject Reference Start Date/Time (SV.SVSTDTC and DM.RFSTDTC, respectively).

The set of visit-site tests can be summarized using a volcano plot to highlight important differences (Figure 4.8 on page 139; reference [19] applies these plots to the analysis of adverse events). Each point represents a statistical test comparing a site versus all others for a particular visit. The y-axis represents the $-\log 10$(raw p-value). To interpret the y-axis of Figure 4.8 on page 139, note that a p-value of

1. 1 equals 0 on the $-\log 10$ scale
2. 0.1 equals 1 on the $-\log 10$ scale
3. 0.01 equals 2 on the $-\log 10$ scale
4. 0.001 equals 3 on the $-\log 10$ scale
5. 0.0001 equals 4 on the $-\log 10$ scale.

The smaller the p-value, the larger the number on the y-axis; y can be thought of as the number of decimal places or number of zeroes in the p-value derived from the comparison of a site to all others for a particular visit. Here, we use the Cochran-Mantel Haenszel (CMH) row mean score statistic to take advantage of the ordinality of study day as described in Chapter 4 of [20]. Further, we apply standardized midrank scores to account for the possibility that the observed study days may not be equally spaced from one another. The x-axis typically represents the value being tested, such as a difference in means or proportions, or in the case of binary data, a risk or odds ratio. Since we are comparing the distribution of study days, we'll use the maximum difference between the suspect site versus the reference (all other sites) across all study days to represent the difference.

For this analysis, or any screening analysis for that matter, it is important to account for multiple comparisons. Given the sheer number of tests being performed, it is highly likely that some statistically significant results are merely due to chance. Multiplicity adjustments can be applied in one of two ways: either the p-value can be made larger or the alpha-level applied to each test made smaller to reduce the likelihood of false-positive outcomes. Multiplicity adjustments are easily applied using a volcano plot; any of the raw p-values at or above the reference line are considered significant when accounting for the False Discovery Rate (FDR) method of Benjamini and Hochberg [21]. Among the rejected null hypotheses from a family or set of multiple tests, the expected proportion of erroneous rejections is defined as the FDR (typically set at 0.05). While this particular method of adjustment typically does not control the overall family-wise error rate (which could lead to false positives), it allows for greater power among the individual comparisons. If stricter multiplicity adjustments were applied, such as the Bonferonni, it would make it extremely difficult to identify any fraud or quality issues. However, in cases where there was a priori knowledge about the performance of a particular site, the number of statistical tests could be limited to this site and stricter multiplicity adjustment could be applied; interested readers can review [22] for appropriate methods.

Figure 4.8 Volcano Plot

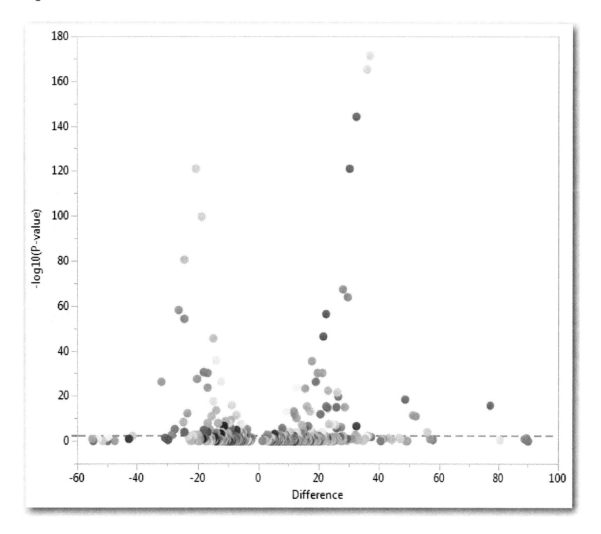

The dialog of **Perfect Scheduled Attendance** has limited options available. First, the analyst can limit the sites considered to be the suspect site based on the number of subjects available from the selected **Analysis Population** on the **Filters** tab (default = 5). Finally, depending on the minimum number of subjects available for analysis per site, asymptotic tests to compare the visit distributions may not be appropriate. A resampling-based MH exact test is available, and the user is free to specify the number of monte-carlo samples for computing p-values (default = 5000). For those interested in additional detail from the SAS documentation, the dialog option for the exact test applies EXACT MHCHI / MC n = 5000 within PROC FREQ.

Assuming sufficient sample size is available for asymptotic tests to be appropriate, click **Run** for **Perfect Scheduled Attendance** to analyze the **Nicardipine** data (Figure 4.9 on page 141). The results are not very dramatic. Most points hover at (0,0), though there is some variability for these points due to very small differences and some jittering that is applied to make it easier to select points. This implies almost no difference between the study day distributions (with corresponding p-values near 1). Only a single site-visit combination appears to be statistically significant, with a p-value of $10^{-12.29}$. Selecting this point and clicking the **Visit Bar Chart** drill down generates Figure 4.10 on page 142 to compare the distribution of Visit 1 with Site 39 as the suspect site. Hovering over the red bar at Study Day -1 tells you that 5.88 percent of these visits occur on this day. Despite the statistical significance, this result hardly seems of interest. The data underlying the bar chart is available for review by going to the red triangle menu, **Script > Data Table Window**. You'll notice that a single visit, out of 17 total visits, occurs on study day -1. In this instance, it is possible for the date associated with Study Day -1 (often the day prior to dosing) to be a data entry error (given that all other values for visit 1 occur on Study Day 1), and this should at least result in a query for the site.

However, the expected sample size for the cells is likely very small, making the choice of an asymptotic test questionable. Statistical significance or not, the evidence available doesn't confirm any wrongdoing or sloppiness on the part of the site. However, in general, if this analysis (or any, for that matter) is to serve as evidence of misconduct, the appropriate analysis should be applied. The exact option could be selected from the dialog to perform the analysis for all site-visit combinations, though this could take considerable time to finish. As another possibility, the data table window that supplies the bar chart can be used directly: Go to **Analyze > Fit Y by X**, set Study Day of Start of Visit as **Y, Response**, Label as **X, Factor** and Frequency Count as **Freq**. The resulting analysis (Figure 4.11 on page 143) shows a p-value that will likely not be significant when applying the FDR adjustment across 530 tests. The number 530 is obtained using **Ctrl-A** to select all points from the volcano plot in Figure 4.9 on page 141 and clicking **Show Sites** to open a data table of 530 rows. From this table, we get a sense that each and every day a subject is in the hospital corresponds to a new visit for the **Nicardipine** study.

Figure 4.9 Perfect Scheduled Attendance Volcano Plot

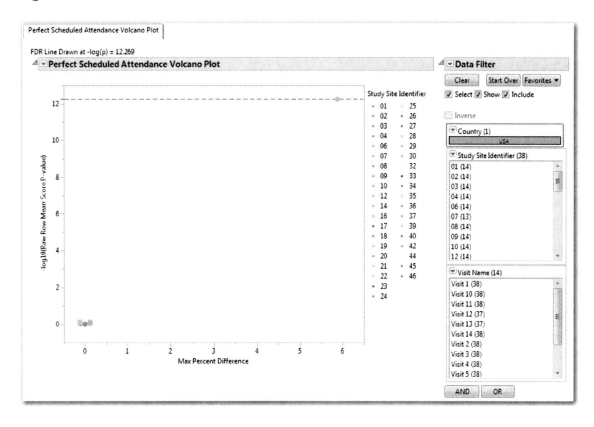

Figure 4.10 Distribution of Study Days for Visit 1 for Suspect Site 39

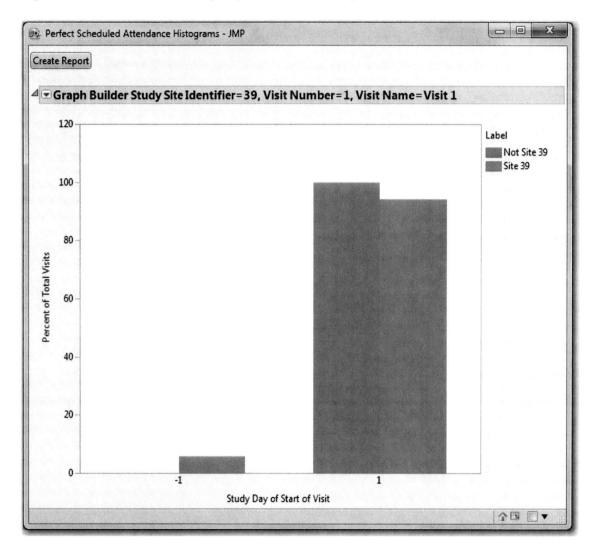

Figure 4.11 Contingency Analysis of Study Days for Visit 1 for Suspect Site 39

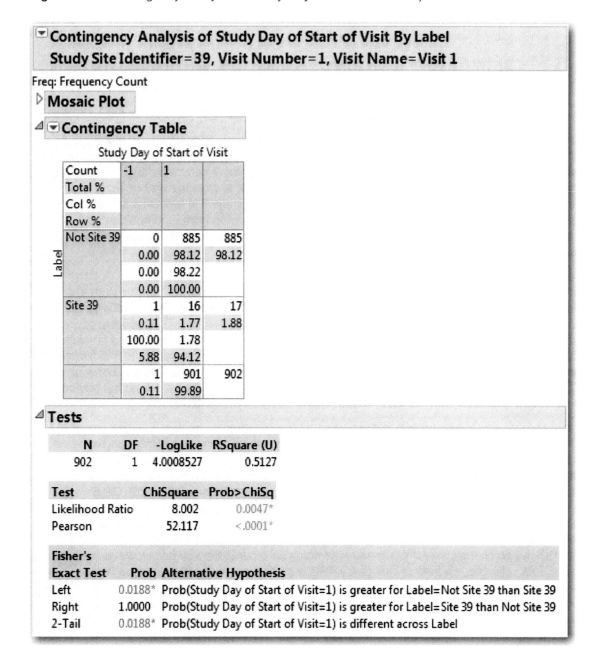

4.3 Measurements Collected at the Clinical Site

4.3.1 Tests with No Variability

Our previous discussion focused on the unusual results that may arise as part of study scheduling. Here, we will focus on the data that is collected at the study site through the various tests, procedures, and examinations performed, or questionnaires completed by the patients. In other words, these data are from the planned evaluations that assess the well-being of the participant, as well as those results that indicate whether the patient is receiving some benefit from the treatment they are receiving. In CDISC, these data come from Findings domains. Some examples of Findings domains include ECG test results (EG), vital signs (VS), laboratory measurements (LB), and physical examinations (PE); there are, of course, many others.

Fraudulent behavior and other severe quality problems are likely limited to a single or small number of clinical sites. Our ability to detect unusual data may depend on how the data at one site may differ from the rest of the pack, as in our analysis of visit schedules in Section 4.2.2. Such signals may include differing response rates or trends across time, or unusual or atypical associations between variables. While it may be straightforward to make up a blood pressure at any given time point, the effects of time and the relationships of other variables to this fabricated measurement are difficult to account for. I describe some comparisons of Findings between sites in Sections 4.3.3 and 4.3.4. However, the methods in Sections 4.3.1 and 4.3.2 do not rely on comparing data across sites, except perhaps in the frequency that hits occur.

In this section, we discuss tests or procedures that are repeatedly measured across time that exhibit no variability. Given the sensitivity of biological systems to various stimuli, such occurrences are extremely unlikely. For example, it is well known that ECG measurements are extremely variable when measured even within very brief time spans; the recommended approach for analysis is to average at least 3 measurements to reduce error [23]. Other measurements, such as intraocular pressure, have a natural diurnal pattern that should be considered in the design and analysis of trials that utilize such measurements (such as trials for glaucoma) [24]. The **Constant Findings** Report looks for all instances where a particular test result shows no change over the course of the study for a particular subject. However, depending on the data collected, the sets of constant values may be examined within location (xx.xxLOC), method (xx.xxMETHOD), position (xx.xxPOS), specimen (xx.xxSPEC), or planned time point (xx.xxTPT). For example, in an ophthalmology study, results within many Findings domains may be reported for the right or left eye, where xx.xxLOC = OD or OS, respectively. Here, the constant tests will be determined separately for each eye, within subject.

For numeric data, finding constant tests is as easy as looking for all variables with a variance (or standard deviation) of zero. For character data, the result of a questionnaire, for example, will have a particular value occur 100% of the time. In the dialog, users can select to analyze numeric results in standard units (xx.xxSTRESN), character results in standard units (xx.xxSTRESC), or results in original units (xx.xxORRES) across all of their available Findings domains. Note that while xx.xxORRES and xx.xxSTRESC are both character variables, the analysis may provide different results. For example, xx.xxORRES may describe a particular abnormality in detail, while xx.xxSTRESC may only be listed as Normal/Abnormal. Figure 4.12 on page 145 shows the analysis of numeric results in standard units for **Nicardipine**.

On the **Laboratory Test Results** tab for Bilirubin, it appears that 45 subjects had no change in their values from the time they entered the trial until the time they exited. Select this part of the histogram and click **Show Subjects** to open a data table of the results; right-click on **Freq of Standard Numeric Findings** and go to **Sort > Ascending**. Based on the last row, Subject 81010 had four instances of Bilirubin reported as 0.00099990003333 mmol/L. While unusual, these results may not necessarily indicate that fraudulent behavior is occurring, particularly since these findings are spread out across numerous sites. However, an explanation is required. Perhaps some laboratory tests show little variability across time, or the particular values for Subject 81010 indicate that a lower limit for the assay has been reached (reasonable, as many of the other subjects have these particular values). This is certainly a place where knowledge and familiarity of the study procedures would be useful to explain these findings. Further, despite the presence of 45 subjects with constant Bilirubin, as shown in Figure 4.13 on page 147, the majority of subjects had repeated values that occurred only twice before they discontinued. Fewer instances of a repeated value may be more plausible and acceptable, though the tolerance for the observed frequency of repeated values will likely vary for each procedure under investigation. As another example, use **Show Subjects** to open a data table of the 12 constant lab tests for Subject 31001 (Figure 4.14 on page 148). While all of the lab test values are repeated only twice, this is a rather large set of repeated measurements for a single subject, which could indicate the presence of a duplicate record (Section 4.3.2).

Figure 4.12 Laboratory Test Results With No Variability

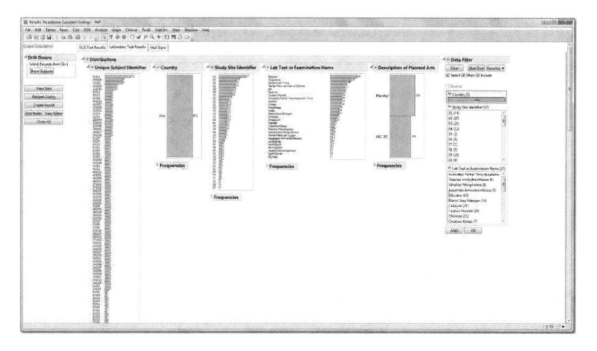

The results above may highlight situations where data are carried forward throughout the CRF by an investigator, identify equipment that is miscalibrated or individuals consistently hitting assay limits, or point to study subjects who are filling out questionnaires or diaries with the sole desire to complete them quickly, rather than accurately. Such results could also highlight certain deficiencies in CRF design on the part of the sponsor. For example, consider a physical exam that is conducted at each study visit. For each body system, the clinician may report whether the exam was normal or abnormal, and if abnormal, list the particular abnormalities. Collecting data in this manner allows the sponsor to summarize abnormality rates across time. However, the abnormalities may be so rare that there are often no results to report; summary tables or listings may describe that all subjects were normal across every visit of the trial. This results in a lot of data to sift through in order to find anything interesting (read: problematic!). Further, if these spontaneously occurring abnormalities are considered clinically significant, it may be necessary to report them as adverse events. Now the study team is in a position of performing consistency checks across CRF domains. Perhaps a more straightforward approach for such exams is to document that they were performed, but report the clinically significant abnormalities as adverse events. This reduces the amount of data collected and eliminates the need for cross-domain consistency checks.

Figure 4.13 Constant Findings for Bilirubin

	Unique Subject Identifier	Study Site Identifier	Description of Planned Arm	Country	Lab Test or Examination Name	Numeric Result/Finding in Standard Units	Freq of Standard Numeric Findings
1	11032	01	Placebo	USA	Bilirubin	0.0053328	2
2	121003	12	NIC .15	USA	Bilirubin	0.0019998001	2
3	121004	12	Placebo	USA	Bilirubin	0.0043329	2
4	122013	12	NIC .15	USA	Bilirubin	0.0032330101	2
5	141021	14	NIC .15	USA	Bilirubin	0.0009999	2
6	161022	16	NIC .15	USA	Bilirubin	0.0009999	2
7	181002	18	NIC .15	USA	Bilirubin	0.0016665	2
8	191001	19	Placebo	USA	Bilirubin	0.0016665	2
9	201002	20	NIC .15	USA	Bilirubin	0.0016665	2
10	201018	20	Placebo	USA	Bilirubin	0.0006666	2
11	21005	02	Placebo	USA	Bilirubin	0.0006666	2
12	21016	02	Placebo	USA	Bilirubin	0.0013332	2
13	21028	02	NIC .15	USA	Bilirubin	0.0013332	2
14	282016	28	Placebo	USA	Bilirubin	0.0009999	2
15	282030	28	Placebo	USA	Bilirubin	0.0013332	2
16	282044	28	NIC .15	USA	Bilirubin	0.0006666	2
17	282063	28	Placebo	USA	Bilirubin	0.0016665	2
18	31008	03	NIC .15	USA	Bilirubin	0.0013332	2
19	31011	03	Placebo	USA	Bilirubin	0.0006666	2
20	31018	03	Placebo	USA	Bilirubin	0.0019998001	2
21	31022	03	Placebo	USA	Bilirubin	0.0009999	2
22	392008	39	NIC .15	USA	Bilirubin	0.001169883	2
23	401001	40	Placebo	USA	Bilirubin	0.0016665	2
24	41013	04	NIC .15	USA	Bilirubin	0.0016665	2
25	41025	04	NIC .15	USA	Bilirubin	0.0016665	2
26	452011	45	NIC .15	USA	Bilirubin	0.002729727	2
27	61004	06	NIC .15	USA	Bilirubin	0.0016665	2
28	81022	08	NIC .15	USA	Bilirubin	0.0013332	2
29	91015	09	Placebo	USA	Bilirubin	0.0009999	2
30	91021	09	Placebo	USA	Bilirubin	0.0039996	2
31	101012	10	Placebo	USA	Bilirubin	0.0016665	3
32	11042	01	Placebo	USA	Bilirubin	0.0016665	3
33	141022	14	Placebo	USA	Bilirubin	0.0013332	3
34	141028	14	NIC .15	USA	Bilirubin	0.0009999	3
35	141031	14	Placebo	USA	Bilirubin	0.0016665	3
36	141055	14	Placebo	USA	Bilirubin	0.0016665	3
37	221012	22	Placebo	USA	Bilirubin	0.0009999	3
38	231020	23	Placebo	USA	Bilirubin	0.0009999	3
39	282024	28	Placebo	USA	Bilirubin	0.0009999	3
40	282042	28	Placebo	USA	Bilirubin	0.0009999	3
41	291004	29	NIC .15	USA	Bilirubin	0.0013332	3
42	322001	32	NIC .15	USA	Bilirubin	0.0009999	3
43	441008	44	Placebo	USA	Bilirubin	0.0013648635	3
44	461013	46	Placebo	USA	Bilirubin	0.0013332	3
45	81010	08	NIC .15	USA	Bilirubin	0.0009999	4

Figure 4.14 Constant Lab Tests for Subject 31001

	Unique Subject Identifier	Study Site Identifier	Description of Planned Arm	Country	Lab Test or Examination Name	Numeric Result/Finding in Standard Units	Freq of Standard Numeric Findings
1	31011	03	Placebo	USA	Alkaline Phosphatase	55	2
2	31011	03	Placebo	USA	Aspartate Aminotransferase	11	2
3	31011	03	Placebo	USA	Bilirubin	0.0006666	2
4	31011	03	Placebo	USA	Calcium	2.125	2
5	31011	03	Placebo	USA	Creatine Kinase	69	2
6	31011	03	Placebo	USA	Creatinine	0.0707200009	2
7	31011	03	Placebo	USA	Partial Pressure Carbon Dioxide	4921	2
8	31011	03	Placebo	USA	Partial Pressure Oxygen	14364	2
9	31011	03	Placebo	USA	Phosphate	1.2597000323	2
10	31011	03	Placebo	USA	Protein	6.5999999	2
11	31011	03	Placebo	USA	Urate	0.1357	2
12	31011	03	Placebo	USA	pH	7.46	2

4.3.2 Duplicate Sets of Measurements

In the previous section, we identified tests or procedures that exhibit no variability within a patient over the course of the clinical trial. In this section, we look for sets of tests where all of the values are repeated in at least two different instances within the clinical site using the **Duplicate Records** report.. What does the term "record" mean? For our purposes, record will refer to a set of tests reported at the same visit and time (if applicable) number within the same CDISC Findings domain, and this often encompasses a single panel on a CRF page. For example, a set of vital sign measurements (stored in the SDTM VS domain) may be collected for a subject (Table 4.1 on page 148).

Table 4.1 Example Duplicate Record for Vital Signs

Visit	Systolic	Diastolic	Heart Rate
Visit 1	140 mmHg	80 mmHg	84 bpm

Given that we expect some variability among the repeated measures for any given variable, it would be extremely unusual and unlikely to find the identical set of measurements within the subject or across subjects within the same clinical site. Multiple hits between two subjects can possibly indicate that CRF pages were copied between these subjects. Even worse, a large number of multiple hits may identify a fictitious subject. Hits within a subject can identify "carried-forward" or data propagation problems similar to those described above in the previous section (e.g., if an investigator forgets to perform a test associated with a particular CRF panel), or they may indicate one or more visits that did not actually occur for the subject, particularly if there are duplicates across several domains for a specific set of visits [18].

The **Duplicate Records** Report allows users to analyze numeric results in standard units (xx.xxSTRESN), character results in standard units (xx.xxSTRESC), or results in original units (xx.xxORRES) across all of their available Findings domains. If selected, two options reduce the number of sets or values within sets reported. **Ignore duplicate records within subject** will delete multiple occurrences of the same subject within each set of duplicate records. If a set of duplicates is based entirely on one subject, these sets are removed. **Ignore duplicate records when covariates don't match** will limit sets of duplicate records where visit number (xx.VISITNUM), location (xx.xxLOC), method (xx.xxMETHOD), position (xx.xxPOS), specimen (xx.xxSPEC), and planned time point (xx.xxTPT) match. The latter option is selected by default. Click **Run** to generate an analysis of the **Nicardipine** clinical trial; the **Vital Signs** tab is displayed in Figure 4.15 on page 149.

Figure 4.15 Distributions of Duplicate Records for Vital Signs

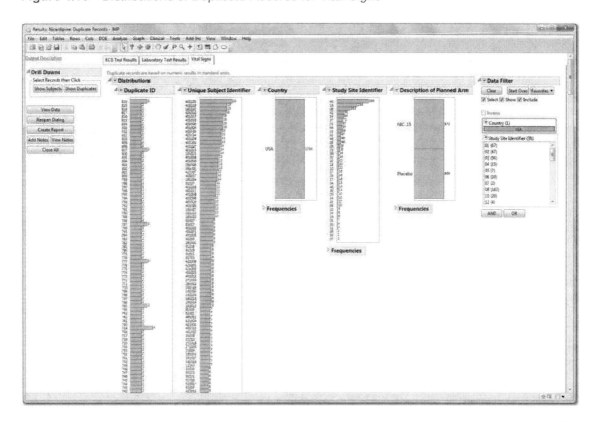

Figure 4.15 on page 149This figure shows a large number of sets of systolic, diastolic, and heart rates across many sites. For example, site 40 has 399 records that contribute to sets of duplicate records. Using the **Data Filter** to subset the histograms to this site and opening the **Frequencies** outline box on the Duplicate ID histogram shows that these 399 records are contained within 185 different sets of duplicates. The Duplicate ID is a system-generated identification value used to refer to sets of unique measurements. **Clear** the **Data Filter**, go to the red triangle of Duplicate ID, and click **Order By > Count Ascending**. Click on Duplicate ID 243 and click **Show Duplicates** (Figure 4.16 on page 150). Notice that rows two through six are for the same subject at Visit 1 with the first set measured at 12:30 and the last set measured at 17:00. We've identified five sets of measurements with the same values recorded over a 4.5-hour period.

Figure 4.16 Vital Signs for Duplicate ID 243

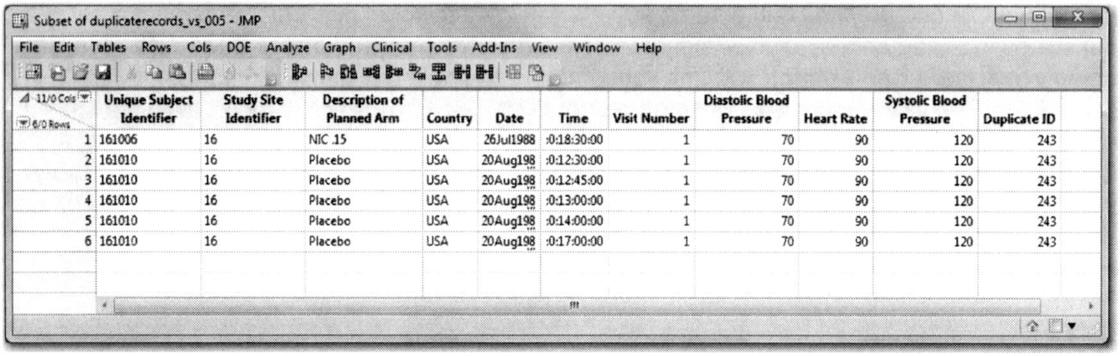

	Unique Subject Identifier	Study Site Identifier	Description of Planned Arm	Country	Date	Time	Visit Number	Diastolic Blood Pressure	Heart Rate	Systolic Blood Pressure	Duplicate ID
1	161006	16	NIC .15	USA	26Jul1988	:0:18:30:00	1	70	90	120	243
2	161010	16	Placebo	USA	20Aug198	:0:12:30:00	1	70	90	120	243
3	161010	16	Placebo	USA	20Aug198	:0:12:45:00	1	70	90	120	243
4	161010	16	Placebo	USA	20Aug198	:0:13:00:00	1	70	90	120	243
5	161010	16	Placebo	USA	20Aug198	:0:14:00:00	1	70	90	120	243
6	161010	16	Placebo	USA	20Aug198	:0:17:00:00	1	70	90	120	243

As another example, select patient 401029 and click **Show Duplicates** (Figure 4.17 on page 151); the data table has 43 records, since clicking **Show Duplicates** will open all records from the Duplicate IDs to which this particular subject contributes. If it is of interest to review only those records for subject 401029, click **Show Subjects** instead. In this case, some records may not have any duplications when subset to this particular individual. To see the Duplicate IDs where this subject has multiple records, add Unique Subject Identifier to the **Data Filter** and subset to this subject. Duplicate IDs with multiple records will show the instances of within-subject duplication for 401029.

Figure 4.17 Duplicate Vital Signs Involving Patient 401029

Unique Subject Identifier	Study Site Identifier	Description of Planned Arm	Country	Date	Time	Visit Number	Diastolic Blood Pressure	Heart Rate	Systolic Blood Pressure	Duplicate ID
1 401009	40	NIC .15	USA	17Oct1988	:0:22:00:00	1	65	75	140	561
2 401029	40	Placebo	USA	07Jun1989	:0:18:00:00	1	65	75	140	561
3 401018	40	Placebo	USA	21Feb1989	:0:18:00:00	1	70	65	145	571
4 401029	40	Placebo	USA	07Jun1989	:0:14:15:00	1	70	65	145	571
5 401029	40	Placebo	USA	07Jun1989	:0:14:30:00	1	70	85	140	575
6 401029	40	Placebo	USA	07Jun1989	:0:19:00:00	1	70	85	140	575
7 401029	40	Placebo	USA	07Jun1989	:0:15:00:00	1	70	85	145	576
8 401029	40	Placebo	USA	07Jun1989	:0:20:00:00	1	70	85	145	576
9 401029	40	Placebo	USA	07Jun1989	:0:16:00:00	1	70	95	150	578
10 401035	40	Placebo	USA	15Aug198	:0:20:00:00	1	70	95	150	578
11 401020	40	NIC .15	USA	08Mar1989	:0:16:00:00	1	80	85	130	584
12 401029	40	Placebo	USA	07Jun1989	:0:17:00:00	1	80	85	130	584
13 401006	40	NIC .15	USA	22Sep1988	:0:11:00:00	4	80	80	140	632
14 401018	40	Placebo	USA	24Feb1989	:0:14:00:00	4	80	80	140	632
15 401029	40	Placebo	USA	10Jun1989	:0:14:00:00	4	80	80	140	632
16 401029	40	Placebo	USA	11Jun1989	:0:18:00:00	5	60	65	105	633
17 401029	40	Placebo	USA	11Jun1989	:0:22:00:00	5	60	65	105	633
18 401012	40	Placebo	USA	21Nov198	:0:14:00:00	5	60	65	110	634
19 401029	40	Placebo	USA	11Jun1989	:0:06:00:00	5	60	65	110	634
20 401029	40	Placebo	USA	11Jun1989	:0:14:00:00	5	70	80	110	638
21 401033	40	NIC .15	USA	21Jul1989	:0:12:30:00	5	70	80	110	638
22 401009	40	NIC .15	USA	22Oct1988	:0:14:00:00	6	70	72	120	650
23 401029	40	Placebo	USA	12Jun1989	:0:18:00:00	6	70	72	120	650
24 401029	40	Placebo	USA	13Jun1989	:0:18:00:00	7	60	75	100	659
25 401029	40	Placebo	USA	13Jun1989	:0:22:00:00	7	60	75	100	659
26 401029	40	Placebo	USA	14Jun1989	:0:14:00:00	8	70	80	120	676
27 401030	40	NIC .15	USA	19Jun1989	:0:06:00:00	8	70	80	120	676
28 401030	40	NIC .15	USA	19Jun1989	:0:18:00:00	8	70	80	120	676
29 401027	40	NIC .15	USA	13May198	:0:06:00:00	8	80	60	130	678
30 401029	40	Placebo	USA	14Jun1989	:0:06:00:00	8	80	60	130	678
31 401019	40	NIC .15	USA	02Mar1989	:0:18:00:00	8	80	85	120	682
32 401029	40	Placebo	USA	14Jun1989	:0:18:00:00	8	80	85	120	682
33 401012	40	Placebo	USA	26Nov198	:0:14:00:00	10	80	75	130	697
34 401018	40	Placebo	USA	02Mar1989	:0:18:00:00	10	80	75	130	697
35 401019	40	NIC .15	USA	04Mar1989	:0:10:00:00	10	80	75	130	697
36 401029	40	NIC .15	USA	04Mar1989	:0:22:00:00	10	80	75	130	697
37 401029	40	Placebo	USA	16Jun1989	:0:06:00:00	10	80	75	130	697
38 401029	40	Placebo	USA	17Jun1989	:0:06:00:00	11	70	75	120	702
39 401035	40	Placebo	USA	25Aug198	:0:17:30:00	11	70	75	120	702
40 401029	40	Placebo	USA	19Jun1989	:0:14:00:00	13	70	78	115	713
41 401030	40	NIC .15	USA	24Jun1989	:0:06:00:00	13	70	78	115	713
42 401010	40	NIC .15	USA	18Nov198	:0:23:30:00	13	80	80	130	715
43 401029	40	Placebo	USA	19Jun1989	:0:22:00:00	13	80	80	130	715

A large number of duplications between any two (or small sets of) subjects may indicate pages that were copied across patients. While it is possible to review the data table of Figure 4.17 on page 151 to get a sense as to whether this occurred, it is much easier to examine graphically. From the data table, click **Graph > Graph Builder**. Drag Systolic Blood Pressure to **Y**, Visit Number to **X** (be sure to right-click on the blue symbol next to Visit Number and change to Ordinal), and Unique Subject Identifier to **Group X**. Click on the **Smoother** in the Icon Bar. Drag Diastolic Blood Pressure to the lower part of the **Y** axis (Figure 4.18 on page 152). Then do the same for Heart Rate. Click **Done** in the **Control Panel** to get Figure 4.19 on page 153. Comparing the points and the non-parametric curves between the subjects for the various parameters is a straightforward way to identify any sets of subjects with several shared

152 Chapter 4 / Detecting Fraud at the Clinical Site

duplications. Judging by Figure 4.19 on page 153, there do not appear to be any subjects that share numerous similarities with subject 401029 for vital sign measurements.

Figure 4.18 Scatter Plots of Blood Pressures from Duplicate Vital Signs Involving Patient 401029

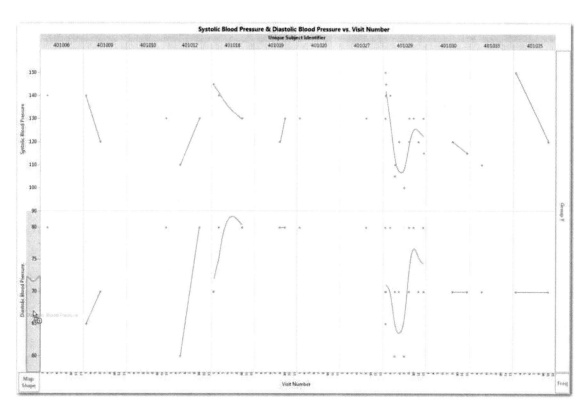

Figure 4.19 *Scatter Plots of Duplicate Vital Signs Involving Patient 401029*

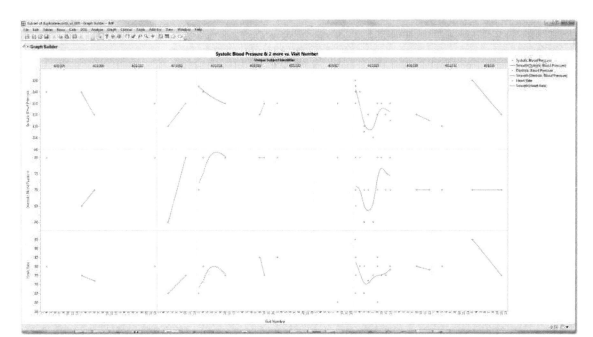

While duplicate records are unusual, it may be possible for the same record to occur between subjects or within a subject. Data that tend to show little variability over time may benefit from a different means of collection (see the discussion on physical exam abnormalities in the previous section) or frequency of collection, such as once per visit or once per study (e.g., how likely is it that height will change over the course of a clinical trial in adults?). However, if a record is repeated among multiple subjects within a clinical site or repeated across multiple visits or time points within a patient, then this would certainly warrant additional investigation, particularly if there are a large number of tests that make up the duplicated record such as a lab panel of a dozen or more tests, or if the data within a duplicate record is expected to be highly variable (such as ECGs). Note that the test set is determined from all tests performed within a Findings domain. This means that if a particular test or tests are not measured at a particular time or visit, the value for those tests will show up as missing within a duplicate set (should any exist). This fact may help explain why there are several duplicates present on the **Laboratory Test Results** tab. Examining the records for subject 321040, who contributes 8 records to 4 Duplicate IDs, shows that all lab test results except for hematocrit are missing. In other words, the Duplicates for this individual are based solely on non-missing hematocrit values. These findings may raise the question as to whether it is likely that only hematocrit values would be reported from a blood draw taken at these time points. Situations such as these may identify quality issues in the reporting of incomplete test results from a site or vendor, or problems in the rows of the data set that inadvertently splits up sets of measurements into unintended smaller sets. Finally, do not convince yourself into thinking that a Duplicate ID with fewer repeats is somehow a less significant finding. For example, **Run** the **Duplicate Records** report with the **Ignore duplicate records when covariates don't match** option unchecked. Click **Show Duplicates** for Duplicate ID 12 to see a set of 26 lab tests repeated exactly one day apart (Figure 4.20 on page 154).

Figure 4.20 Duplicate Laboratory Test Results

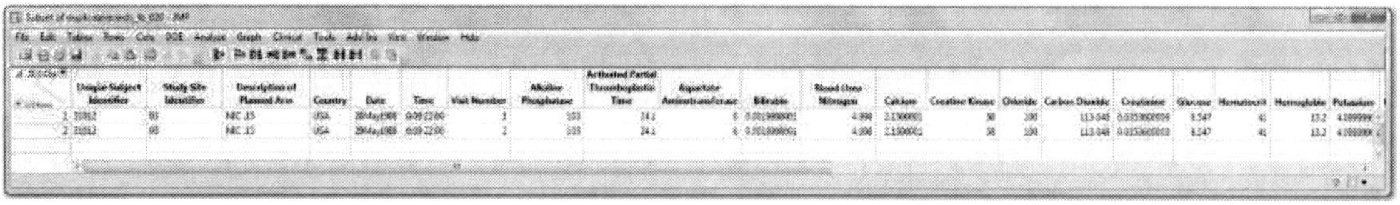

4.3.3 Digit Preference

We return to making comparisons across clinical sites in order to identify quality issues; our interest here is to identify anomalies through tests of the trailing digit for all procedures that provide numeric outcomes. This digit preference analysis is performed in a very similar fashion to the visit schedule analyses of Section 4.2.2. In the absence of evidence to narrow the search to a subset of sites, we can screen for unusual site-test combinations by treating each center as the "suspect site" in turn and comparing the distribution of the trailing digit (i.e., the last digit) to that of all other sites combined. In general, i sites with j procedures performed will result in up to $i \times j$ comparisons (allowing for the possibility that not all procedures were performed at every site). What would we expect to find using such an analysis? In fact, digit preference can identify numerous problems at a study site. Examples include cases where investigators may tend to round manually interpreted or obtained values, instances where diagnostic equipment may be miscalibrated, use of techniques that vary from those specified in the protocol (e.g., taking blood pressure measurements manually rather than by machine), or important differences in subjective measurements, such as the investigator's assessment of clinical signs using a Likert scale (which may suggest training is needed).

Like the **Perfect Scheduled Attendance** analysis of "4.2.2 Study Scheduling" on page 136, results are initially summarized using a volcano plot, allowing the user to select interesting points for further examination. The dialog of **Digit Preference** has only one option available: the analyst can limit the sites considered to be the suspect site based on the number of subjects available from the selected Analysis Population on the Filters tab (default = 5). In the current implementation, tests from all visits and time points are used to define the digit distribution. However, the presence of at least one variable from among location (xx.xxLOC), method (xx.xxMETHOD), position (xx.xxPOS), specimen (xx.xxSPEC), or planned time point (xx.xxTPT) will perform comparisons of trailing digits within the cross-classification of the available variables. The Cochran-Mantel Haenszel (CMH) row mean score statistic is used to take advantage of the ordinality of last digit as described in Chapter 4 of [20]. Further, we apply standardized midrank scores to account for the possibility that the observed last digits may not be equally spaced from one another. Here, the x-axis represents the maximum difference between the suspect site versus the reference (all other sites) across all observed digits to represent the Max Percent Difference. **Run Digit Preference** for **Nicardipine** with the minimum number of subjects equal to five (Figure 4.21 on page 155).

Figure 4.21 Digit Preference Volcano Plot

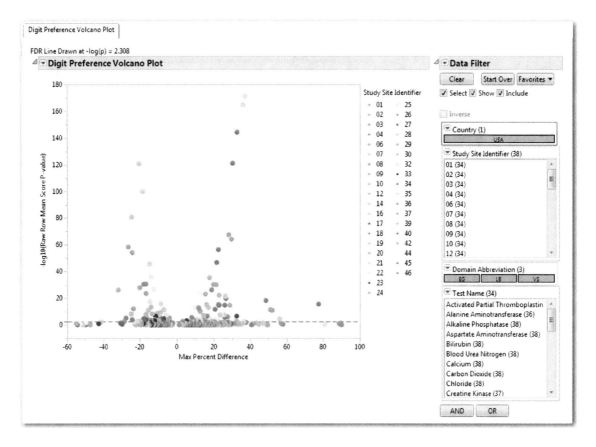

Markers that trend toward the upper corners are those where large statistically significant differences occur. For example, selecting the two light blue markers at the top and clicking **Digit Bar Charts** creates the plot in Figure 4.22 on page 156. Notice that Site 16 reports blood pressure (both systolic and diastolic) with a trailing digit of zero about 35% more often than the other sites. In fact, upon closer examination, no odd last digits are reported at this site at all. Similar findings are available for Site 40 (the two brown markers near the top), though here most of the trailing digits are reported as zero with five reported for a majority of the remaining digits. As was done in Section 4.2.1, we could animate the **Data Filter** to cycle through and subset the figure to clinical sites or domains to more easily review results, or find areas in general need of training. Markers extreme on the x-axis, though near zero along the y-axis, represent large differences between the digit distributions. In these cases, however, there is likely an insufficient sample size to obtain statistical significance (the data table for each bar chart can be surfaced through the red triangle, **Script > Data Table Window**). For certain tests, such as those obtained using a piece of diagnostic equipment that requires periodic calibration, the statistical significance may be of secondary concern. Any large differences in distribution should be examined to determine if the observed results are potentially worrisome.

156 Chapter 4 / Detecting Fraud at the Clinical Site

Figure 4.22 Distribution of Trailing Digits for Blood Pressure Results

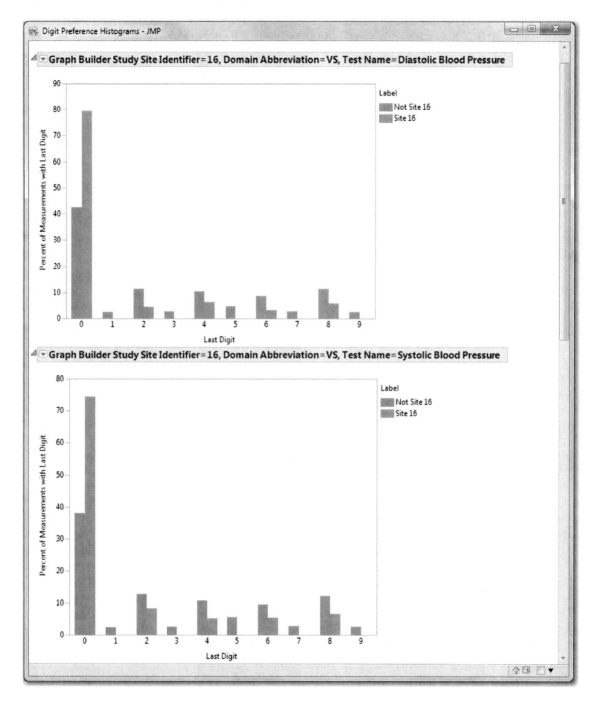

4.3.4 A Brief Interlude from the Fraud Detection Menu

4.3.4.1 Time Trends

In this section, we make use of **Findings > Findings Time Trends** to present the results of various tests, procedures, and examinations by site over the course of the clinical trial. Such figures can help analysts identify outliers and unusual trends or patterns in the data that are collected across time (functions of xx.xxSTRESN by xx.VISITNUM). Coupled with the knowledge of how these data should behave while on treatment or with worsening disease, these line plots can help identify specific points, patients, or sites that require further investigation to understand the unusual results. Although the **Findings Time Trends** Report has numerous options available, we highlight only a fraction of them here. Interested users can go to **Clinical > Documentation > Help > User Guide** or **Help > Books > JMP Life Sciences User Guide** for more information on the various options.

For the **Nicardipine** study, select VS as the **Findings Domain to Analyze**, change **Treatment or Comparison Variable to Use** to **Specified Below**, then select SITEID and click the arrow (-->) button to select it as the comparison variable. Click **Run**. When the results window opens, click **View Data**, right-click on Study Site Identifier in the **Columns** menu in the left panel or in the data table, and select **Label/Unlabel**. Click on the X in the upper right of the data table window, and when prompted, click **Hide Data Table**. This exercise labels points in the figure with Study Site Identifier for ease of identification when the analyst hovers with the cursor. Figure 4.23 on page 158 presents systolic and diastolic blood pressure averaged by site across visits from the **VS Treatment Time Trends** Tab. The second Tab presents time trends of blood pressures by subject.

158 Chapter 4 / Detecting Fraud at the Clinical Site

Figure 4.23 Time Trends of Systolic and Diastolic Blood Pressures by Study Site Identifier

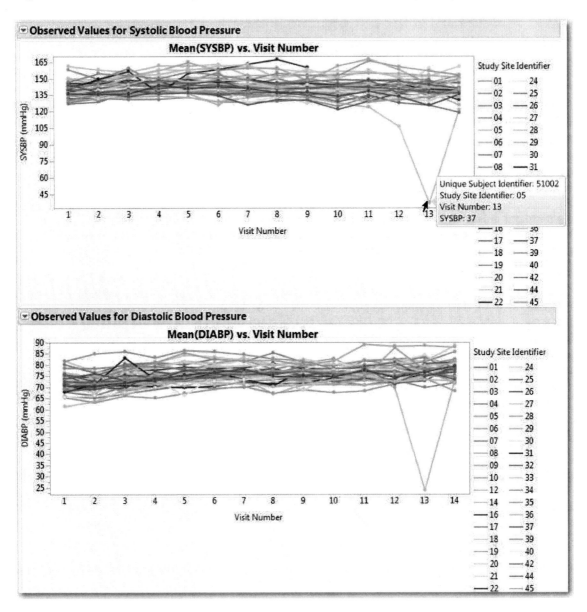

In Figure 4.23 on page 158, we can easily attribute the very large dip for blood pressures at Visit 13 to Site 05. Though the label presents subject 51002 as the individual responsible for this outcome, this can be somewhat misleading. Because both Tabs use the same data table, JMP computes the summary statistics (here, the means) directly, based on the selected option within **Graph Builder** (the platform used to create each figure) and using Study Site Identifier as an overlay on **VS Treatment Time Trends** Tab. To identify which subjects may contribute to this unusual finding, switch to the **VS Subject Time Trends Tab** and use the **Data Filter** to subset to data for site 05. Right-click in the plots and go **Add > Points**; hovering over the points provides the informative labels needed to identify the subject, which in this case turns out to be subject 51002 (Figure 4.24 on page 159). Selecting the line (not individual points) for this subject and going to **Table > Subset > OK** will subset the full data table to the records for this particular subject (Figure 4.25 on page 160, uses **Cols > Reorder Columns** for death information). Given the precipitous drop for both blood pressures at the same visit and the fact that the patient died on this day (based on death date and the first date of dosing), the blood pressure readings at Visit 13 are likely accurately reported. In the absence of such information, the site should at minimum be queried for correctness of the findings.

Figure 4.24 Time Trends of Blood Pressures for Subjects from Clinical Site 05

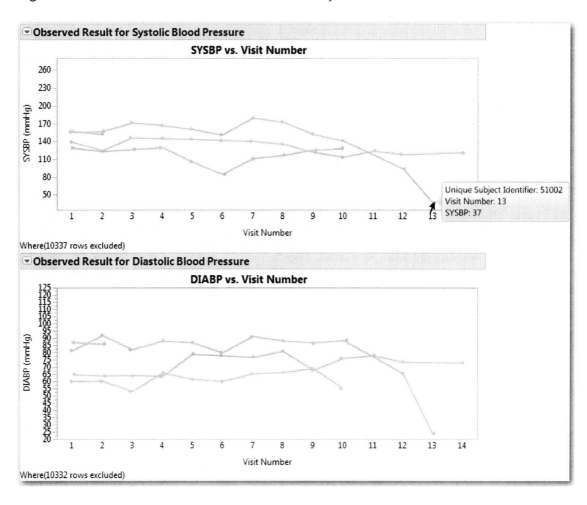

Figure 4.25 Vital Signs for Subject 51002

Unique Subject Identifier	Visit Number	SYSBP	DIABP	HR	Patient Died Flag	Date of death
51002	1	155.57142857	81.571428571	66.785714286	Y	13Nov1988
51002	2	157.14285714	91.857142857	84.285714286	Y	13Nov1988
51002	3	171.5	82.25	83.75	Y	13Nov1988
51002	4	167.33333333	88	95	Y	13Nov1988
51002	5	160.5	87	84.75	Y	13Nov1988
51002	6	151.2	80	88.4	Y	13Nov1988
51002	7	179.75	91.25	79	Y	13Nov1988
51002	8	173	88.25	95.5	Y	13Nov1988
51002	9	152	86.75	109.75	Y	13Nov1988
51002	10	141.25	88.5	103.25	Y	13Nov1988
51002	12	94.5	65.75	75.5	Y	13Nov1988
51002	13	37	24	40.5	Y	13Nov1988

The previous analysis summarizes data for all treatment groups combined, since the goal is to identify unusual results at the site level. For endpoints where it is expected that the treatment will have little impact, such an approach is likely acceptable. However, for those endpoints expected to be impacted by drug application, plotting the results by treatment is a good idea. To do this, users can select ARM for **By Variables** on the **Filters** Tab of the dialog. For example, Figure 4.26 on page 161 shows the results for sites by Description of Planned Arm for diastolic blood pressure; Site 34 appears to have higher diastolic blood pressure compared to other sites for the Nicardipine arm, particularly for the earlier visits. Another option is to use **Data Filter** animation to compare treatments by site, animating the filter to subset by Study Site Identifier. Either approach may prove extremely enlightening, particularly for studies where study personnel have knowledge of the treatment assignments (e.g., single-blind or open-label studies).

Figure 4.26 Time Trends of Blood Pressures by Study Site Identifier and Treatment

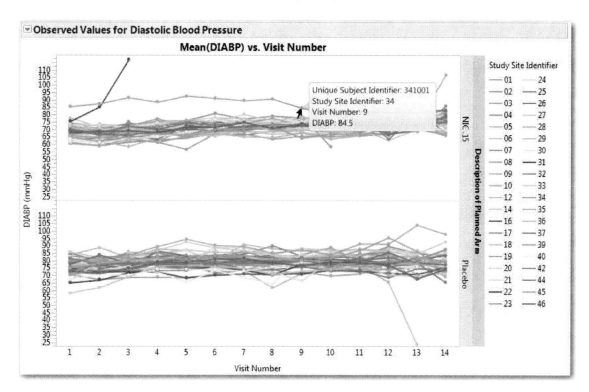

4.3.4.2 Plots of Summary Statistics by Site

Using the data table generated from **Findings > Findings Time Trends** in the previous section, we can create some very straightforward summary graphics to compare results across sites. Recall that in the introduction to this chapter, I described an example where a fellow statistician identified some inappropriate behavior at a site because the variability of the measurements for a particular reader was too low compared to other site personnel. While we may be able to easily fabricate values around a mean, mimicking higher-order summary statistics for a variable can be very difficult without computer assistance [1,17]. For the **Nicardipine** study, select VS as the **Findings Domain to Analyze**, change **Treatment or Comparison Variable to Use** to **Specified Below**, then select SITEID and click the arrow (-->) button to select it as the comparison variable. Click **Run**. Open the underlying data table by clicking **View Data**, then go to **Graph > Graph Builder**. Drag SYSBP to **Y** and Study Site Identifier to **X**. Click **Done**. Figure 4.27 on page 162 presents a box plot summarizing systolic blood pressure measurements across all visits by site. This plot can be used to compare the distributions of blood pressure across sites, and can easily highlight any severe departures in mean, median, or variability, as well as highlight any outliers that may be present. To present results by study visits, drag Visit Number to **Group X** (as done below). Further, if needed, we can go to the red triangle and click **Script > Local Data Filter** in order to subset the results to subjects meeting specific demographic or findings criteria for subjects, or subset to particular visits for review. Box plots are also available from **Findings > Box Plots**. To present results by study visits, drag Visit Number to **Group X** (as done below).

162 Chapter 4 / Detecting Fraud at the Clinical Site

Figure 4.27 Box Plots of Systolic Blood Pressure by Study Site Identifier

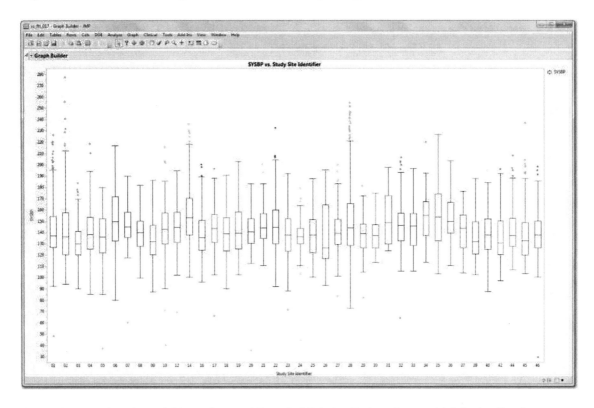

Go to the red triangle and **Show Control Panel**. Click the **Points** icon, which is the first icon above the **Graph Builder** area, select **Variance** as the **Summary Statistic**, right-click on the plot, and choose **Graph > Marker Size > XXXL** to make the symbol markers larger. Drag Visit Number to **Group X** so that sites are presented within each visit. Click **Done** (Figure 4.28 on page 163). This figure can be used to identify particular site-visit combinations where the variability appears to be too small or too large for our liking when compared to other sites. Alternatively, we can drag Description of Planned Treatment to **Overlay** in order to separate computations of variance based on treatment (Figure 4.29 on page 164). This may be particularly important for those endpoints likely to be affected by study therapy. Here we see the site-visit-treatment combinations associated with higher variability tend to come from the Nicardipine treatment arm. Choosing the **Mean** as the **Summary Statistic** can help identify site-visit combinations with unusual levels of response.

Figure 4.28 Variance of Systolic Blood Pressure by Study Site Identifier and Visit Number

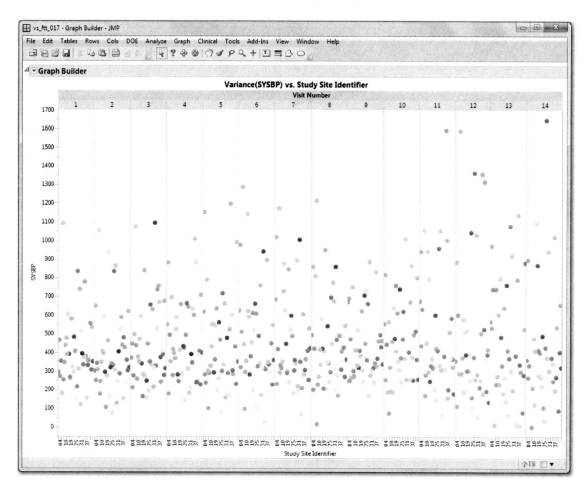

Figure 4.29 Systolic Blood Pressure Variance by Study Site, Visit Number and Treatment

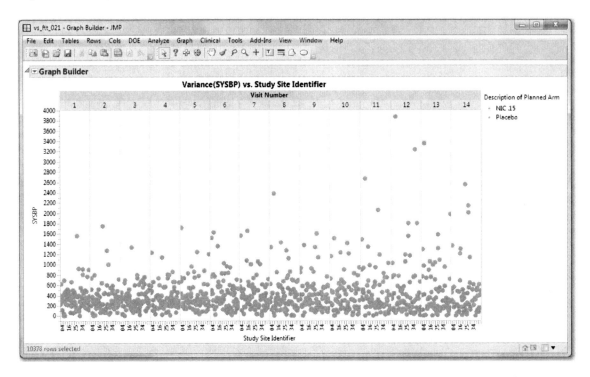

4.4 Multivariate Analyses

4.4.1 Multivariate Inliers and Outliers

In statistics, most of us are familiar with the term "outlier." Merriam-Webster defines outlier as "a statistical observation that is markedly different in value from the others of the sample." In a box plot, an outlier may be identified as a point that exceeds 1.5 times the interquartile range (the distance between first and third quartiles) beyond the first or third quartiles. On the other hand, an "inlier" is a value that lies close to the mean. In a univariate setting, a value close to the mean would not raise any eyebrows; in fact, this is entirely expected. However, as Evans points out, it would be unlikely for an observation to lie near the mean for a large number of variables [17]. So while outliers may be problematic and should be reviewed for correctness or to assess the safety of trial participants, inliers may be more likely to represent observations that are "too good to be true" or "too good to be real." How better to escape detection than to create individual values that do not immediately stand out?

It is possible to review a histogram or box plot to identify any outliers that may arise for each of the covariates under investigation. This approach, however, does not consider the relationship among the covariates; it may be that two points that look rather innocuous when considered individually may be extremely unusual when considered simultaneously. Extend such thinking to three dimensions and beyond, and it is possible to identify points in the multivariate space that appear unusual, either as outliers or as inliers. Two- and three- dimensional scatter plots can help identify unusual points graphically, but such approaches are not available for higher dimensional problems. However, for the general problem with k covariates, we can make use of Mahalanobis distance, which I first introduced in Section 3.3.1.2 [25]. Mahalanobis distance can be used to calculate the distance between two vectors of data (to assess similarity) or from a vector to a particular point in multivariate space (typically the multivariate mean or centroid). The Mahalanobis distance differs from Euclidian distance in that it accounts for the correlation between the variables, thus considering that points may not be distributed spherically around the centroid. It is straightforward to compute in SAS or is available from **Analyze > Multivariate Methods > Multivariate** in JMP. For our purposes we'll be comparing vectors of subject data to the centroid [26].

The **Multivariate Inliers and Outliers** Report creates a data set comprising one row per subject from a set of CDISC-formatted data sets; the dialog is presented in Figure 4.30 on page 166. Users are free to **Include these data in the analysis**:

- Age and Sex can be obtained from the DM distribution (DM.AGE and DM.SEX, respectively)

- Variables for lab tests, vital signs, symptoms, and other Findings data will be generated by visit number and time point (xx.VISITNUM and xx.xxTPTNUM) and included in the analysis if the Findings domains is checked (using xx.xxSTRESN, the Numeric Result/Finding in Standard Units);

- Variables representing frequencies of medications, tobacco use, and other interventions domains will be included by selecting Interventions domains (xx.xxDECOD for Standardized Names);

- Variables representing frequencies of adverse events, medical history terms, and other event domains will be included by selecting Event domains (xx.xxDECOD for Standardized Names).

Based on dialog options, variables exceeding a certain percentage (default 5%) of missing data across the population of subjects determined by the **Filters** tab will be excluded from the analysis, since subjects with any missing data for the variables of interest cannot have Mahalanobis distance computed.

Figure 4.30 Multivariate Inliers and Outliers Dialog

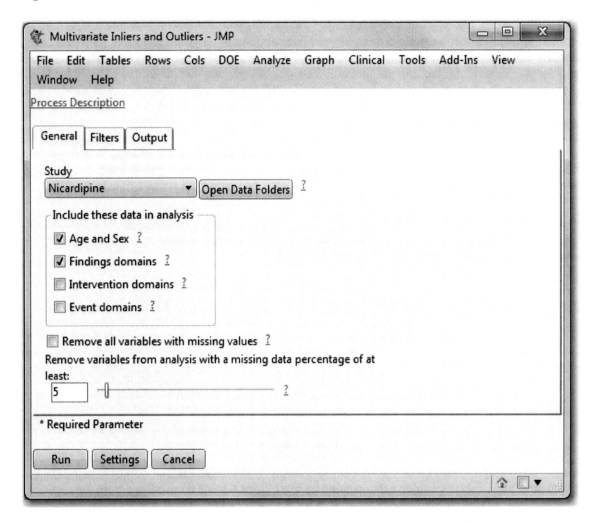

It is important to note that a balanced approach must be taken between using as much data as possible versus the potential to exclude too many patients in your review. Study participants who discontinue early would be expected to have numerous missing values, so it is reasonable to lose some subjects from the analysis when considering the majority of the data. However, selecting **Remove all variables with missing values** excludes all variables with any missing data for the selected population. The resulting data set will then be used to compute Mahalanobis distance.

Run the analysis with the options as selected in Figure 4.30 on page 166; the results are presented in the **Mahalanobis Distance Tab** of Figure 4.31 on page 167 (an analysis generated on the log-scale is not presented). The dashed red line is drawn at the expected value of Mahalanobis distance, which is the square root of the number of covariates used in its computation. The figure shows a box plot of distance measures for each subject compared to the centroid of the multivariate distribution. The distance measure is computed from the numeric results in standard units (xx.xxSTRESN) of 17 of 468 possible variables derived from vital signs, laboratory measurements, and ECGs across the many visits of the trial. The remaining variables had missing data rates exceeding 5% and were removed from analysis, and these variables and the frequency of missing data are described on the **Missing Data** Tab. Fifty-two subjects did not have distance measures computed due to 15 variables containing some missing data. This figure can identify outliers (large distances) or inliers (small distances). Of course, a Mahalanobis distance of zero would represent a subject that lies on the mean for every variable.

Figure 4.31   Box Plots of Mahalanobis Distance

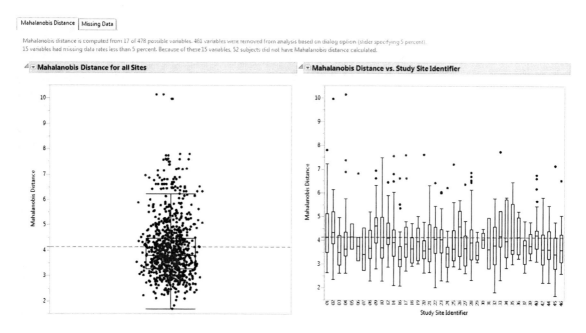

Figure 4.31 on page 167 also presents a box plot of Mahalanobis distance by study site to identify if any particular site is extreme. This plot may help uncover any possible outliers or data errors among the analyzed variables at each study site, but it also may describe key differences in study population across the sites. The length of the box plots indicates how subjects vary within a given site around the multivariate mean. Large variability can reflect a diverse study population or a site that may benefit from additional training. Low variability may reflect a particularly homogeneous population. In any event, any site that distinguishes itself from the others may require greater scrutiny. For example, we can highlight the two patients with distances around 10 and use **Show Subjects** or **Profile Subjects** to review their individual data points. Alternatively, we can use **Cluster Subjects** to assess the similarity of selected subjects based on the underlying data table of Findings results, or we can **Create Subject Filter** to subset follow-up analyses to the selected subjects of interest. Finally, while the **Multivariate Inliers and Outliers** Report makes every effort to analyze as much data as possible, users can analyze any subset of variables from the underlying data table. Click **View Data** and then **Analyze > Multivariate Methods > Multivariate and Correlations**; select variables for analysis as **Y, Columns**. In the platform output, go to the red triangle and click **Outlier Analysis > Mahalanobis Distance** to generate a plot. For example, Figure 4.32 on page 168 is based on systolic blood pressures only. Other multivariate outlier analyses are possible from the hot spot **Jacknife Distances** or **T Square**. Selecting points from

this plot will highlight points in the data table, and these subjects can be profiled using **Profile Subjects**.

Figure 4.32 *Mahalanobis Distance Calculated From Systolic Blood Pressures*

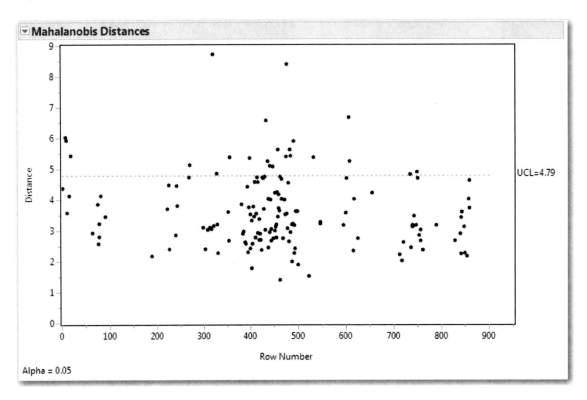

4.4.2 Hierarchical Clustering of Subjects Within Clinical Sites

In Section 4.3.2, we examined a form of data propagation in which sets of measurements with the same values occurred in multiple places throughout the database. In other words, a record or CRF panel was duplicated in two or more places. Finding duplicate records within a subject may suggest data that was carried forward elsewhere in the CRF to avoid performing a set of study procedures. Alternatively, the tests or procedures may have been overlooked, possibly due to some adverse experience that the patient was undergoing at the time. Here, the study personnel may impute data from an earlier visit to correct the oversight. Finding duplicate records between subjects can identify situations where pages were copied from one CRF to another. While these duplications can identify numerous quality issues in the data, there are limitations to this particular analysis tactic. If an investigator or other site personnel is going to engage in such activities, it seems unlikely that he or she would be so careless as to copy pages or records verbatim without modifying some of the values. For example, merely changing a duplicated heart rate by a single beat per minute would prevent the duplicate records check from detecting any wrongdoing (though plots or comparisons of variance may uncover this case). Further, while it is possible to generate frequencies of exact hits between two subjects across several domains, such an approach does not truly assess the similarity of the subjects since the records that differ cannot be directly compared.

In Section 4.4.1, I described how Mahalanobis distance can be used to identify multivariate inliers and outliers when compared to the centroid. Mahalanobis distance can also be used as a measure of similarity between every pair of subjects. To compute Mahalanobis distance, a principal component analysis is performed on a set of covariates; the resulting components are then passed into PROC DISTANCE to calculate the Euclidian distance between pairs of subjects [26]. The Euclidian and Mahalanobis distance are identical in this case, since the principal components are uncorrelated to one another. When any missing data are present for a subject, however, all of the principal components for this subject are missing, which prevents the calculation of any pairwise distances with other subjects. Instead, in this section, we calculate the Euclidian distance (centered and scaled) between subjects since PROC DISTANCE is capable of computing distances in the presence of missing values for covariates (the distance is calculated based on all non-missing pairs).

So why compute similarity measures between subjects in the first place? Our goal in this section is to identify fabricated subjects. Particularly small pairwise distances may suggest that the subjects are slightly modified copies of one another. In the absence of missing data, a pairwise distance of zero would suggest a perfect match; a pairwise distance of zero with some missing data would suggest a perfect match for all of the non-missing values. Here, analyses are limited to pairs of subjects within the same clinical site, hence the name of the report: **Cluster Subjects Within Study Sites**. The dialog is presented in Figure 4.33 on page 170.

Figure 4.33   Cluster Subjects Within Study Sites Dialog

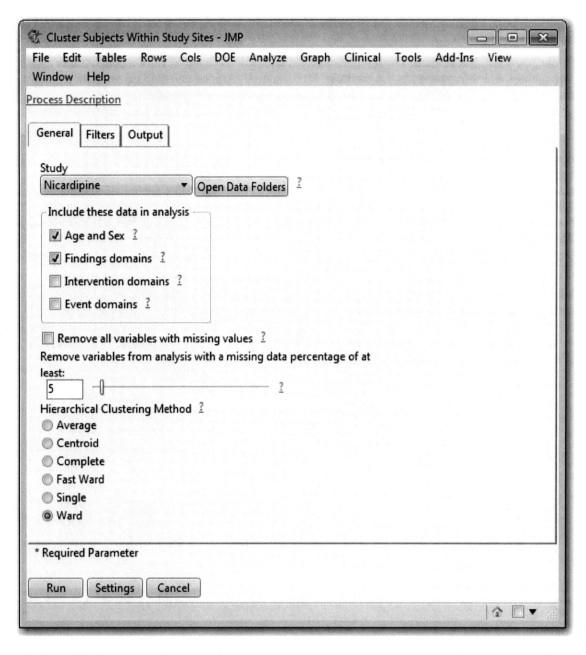

Similar to **Multivariate Inliers and Outliers**, the user may choose to **Include these data in the analysis**:

- Age and Sex can be culled from the DM distribution (DM.AGE and DM.SEX, respectively);

- Variables for lab tests, vital signs, symptoms, and other Findings data will be generated by visit number and time point (xx.VISITNUM and xx.xxTPTNUM) and included in the analysis if the Findings domains is checked (using xx.xxSTRESN, the Numeric Result/Finding in Standard Units);
- Variables representing frequencies of medications, tobacco use, and other interventions domains will be included by selecting Intervention domains (xx.xxDECOD for Standardized Names);
- Variables representing frequencies of adverse events, medical history terms, and other event domains will be included by selecting Event domains (xx.xxDECOD for Standardized Names).

Because we calculate Euclidian distance directly on the covariates, the analysis can accommodate missing values. By default, however, the dialog excludes variables where the percentage of missing values exceeds five percent across the population of subjects determined by the **Filters** tab. Alternatively, any variables with missing data can be removed from the analysis. Finally, users can select a **Hierarchical Clustering Method** (default Ward). For details on the various clustering methods, users are encouraged to review the technical details within the Hierarchical Clustering description within the **Help > Help Contents**. Running the Report with the options selected as in Figure 4.33 on page 170 generates the output in Figure 4.34 on page 171.

Figure 4.34 Box Plots of Between-Subject Distances

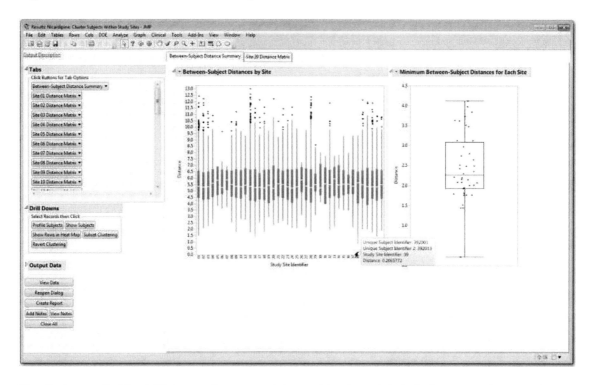

The **Between-Subject Distance Summary** tab provides two plots to help answer the question "How similar is too similar?" Of course, a distance of zero between any two subjects would indicate an exact copy, but analysts need to have some idea of what is reasonable to expect between subjects among the different sites. The left-most figure presents box plots to summarize the distribution of pairwise distances within each study site, while the plot on the right summarizes the distribution of the most similar pair of subjects from each site. This analysis is motivated by a need to identify fabricated subjects, but it also helps describe the study population available at the clinical site. Small or large boxes in the left-most plot indicate centers with a more homogenous or heterogeneous population, respectively. From the plots, we can identify pairs of subjects (392001, 392013) and (392013, 392015) from Site 39 as having the smallest pairwise distance. By default, additional analysis is presented automatically for the site with the minimum pairwise distance (here, the **Site 39 Distance Matrix** tab); similar analyses are available for every site and can be opened by clicking **Open Tab** for any of the **Distance Matrices** under the **Tabs** outline box.

The **Site 39 Distance Matrix** tab presents two figures (Figure 4.35 on page 173). The one at left is a box plot of all pairwise distances for subjects at Site 39. The one at right contains our hierarchical clustering using the Ward method for combining clusters and a heat map to graphically describe the magnitude of the pairwise distances, where the white to red legend indicates less to greater similarity, respectively. From the box plot, we can select the most similar pair of subjects (those closest to zero) and highlight their rows in the heat map by clicking the **Show Rows in Heat Map** drill down (Figure 4.36 on page 174). This can be useful for identifying the cluster membership for any selected pairs of subjects within the **Hierarchical Clustering** analysis.

Figure 4.35 Distance Matrix for Site 39

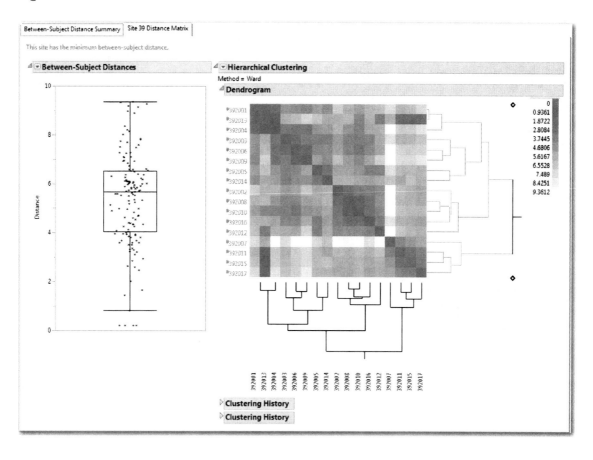

Figure 4.36 Show Rows in Heat Map Drill Down

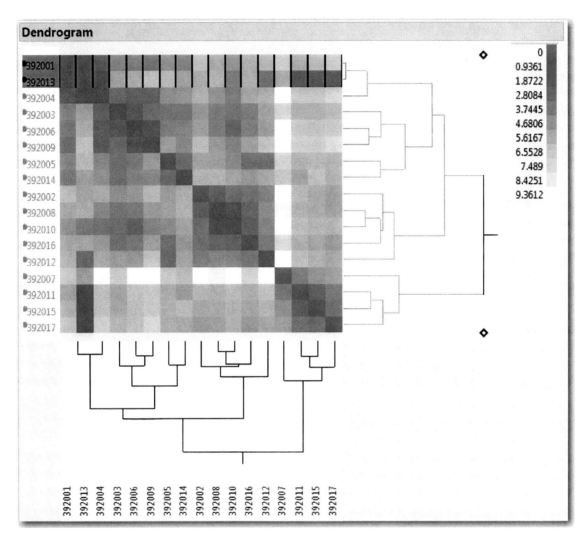

So why bother with clustering? Why not just rely on small pairwise distances? The cluster analysis can help identify sets of subjects of size three or greater that may be overly similar. The analyst may deduce this related triplicate from three separate pairwise differences, but the dendrogram may provide additional insight into the similarity among subjects at the clinical site. Though the pairwise distance for 392001 and 392013 seems very small, the cluster analysis suggests two clusters of subjects at the site (based on the red and blue color of the subject IDs in the dendrogram of Figure 4.36 on page 174). Of course, the clustering may vary based on the selected method; alternate methods can be utilized to examine the robustness of findings. The **Subset Clustering** drill down will subset the cluster analysis to those points selected in the box plot. **Revert Clustering** will include all subjects in the cluster analysis.

Two approaches are available to review data for subjects that appear too similar to one another. The **Show Subjects** drill down opens a data table displaying each subject as a row with columns as variables. Alternatively, **Profile Subjects** can be used to generate graphical displays of all of a subject's data. I illustrate both of these drill downs in Section 5.2.

4.5 Final Thoughts

While I addressed a variety of techniques to identify data anomalies resulting from potential fraud and other quality issues at the clinical sites, the list of methods I presented is by no means complete; in fact, I have only just scratched the surface! The literature contains many additional analyses useful for identifying unusual data arising in a clinical trial [1,3-5,17,18,27-29]. Though these analyses are often presented in the context of detecting issues as a result of misconduct at the clinical site, hopefully the presentation above has illustrated how these analytical techniques can be used to identify any number of quality issues that may arise in a clinical trial. I believe that these techniques should be applied early and often so that the necessary interventions can be applied while the trial is ongoing. Once the database is locked for the primary analysis, the number of options available to the study team to address quality issues is reduced considerably.

However rare site misconduct may be, having a set of easy-to-use analyses and reports to identify potential fraud or quality concerns invalidates worry over the investment in effort. In short, performing such checks of the data is a small price to pay to protect the patient, ensure study integrity, and minimize potential disruption. Further, with the current trend toward centralized monitoring and reduced presence at the clinical sites, such efforts can bring additional peace of mind to the study team and regulatory bodies that the data are of the utmost quality. While the statistical and graphical techniques described are useful for identifying anomalies, note that these techniques by no means represent a magic spell that will identify problem sites, patients, or tests and place them front and center for resolution. These analyses will highlight outliers, unusual trends, and patterns. Experience in clinical research, knowledge of the particular disease area and local regulations, and additional investigation are needed to fully and correctly interpret any findings and work toward an appropriate solution. Further, as O'Kelley has shown, even an analyst armed with the appropriate statistical methods and knowledge of wrong-doing can fail to identify problematic sites [30].

So where do we go from here? Certainly, there are numerous individual methods to identify misconduct and quality issues that may be developed and implemented in future versions of JMP Clinical. Existing methods may benefit from various options to further refine the output. The goal of any new method or option should be to help screen the database for potential problems. For example, while the analysis of correlation structure between real and fabricated data may be important, providing informative initial summaries of clinical sites (such as the volcano plots of Sections 4.2.2 and 4.3.3) will be needed to streamline review [27]. There are, after all, numerous sites and data domains with which to contend. Further, as briefly described at the end of Chapter 3, surfacing informative results from various tools from the **Fraud Detection** menu to the **Risk-Based Monitoring** report will allow the opportunity to apply risk thresholds to additional fraud and quality metrics.

Finally, trial sponsors need to recognize their role in potential fraud and data quality. Recall from "4.3.1 Tests with No Variability" on page 144 that poor CRF design can contribute to a large volume of collected data that is of little value, which can potentially cause inconsistencies with other domains. Simple changes can reduce the volume of data and eliminate the need for tedious cross-domain review. Further, while sponsors should do everything possible to minimize missing data, repeatedly querying a site over missing measurements at a visit may encourage site personnel to impute data from other visits. Better for the sponsor to accept that certain tests may have gone overlooked and allow the statistical team to impute appropriate values if necessary. Other effective methods for reducing potential fraud are to simplify eligibility criteria and reduce the amount of data collected [1]. In the next chapter, I describe ways to identify study participants who have enrolled at two or more clinical sites within the same study.

References

1. Buyse M, George SL, Evans S, Geller NL, Ranstam J, Scherrer B, LeSaffre E, Murray G, Elder L, Hutton J, Colton T, Lachenbruch P & Verma BL. (1999). The role of biostatistics in the prevention, detection and treatment of fraud in clinical trials. *Statistics in Medicine* 18: 3435-3451.

2. White C. (2005). Suspected research fraud: Difficulties of getting at the truth. *British Medical Journal* 331: 281–288.

3. Al-Marzouki SA, Evans S, Marshall T & Roberts I. (2005). Are these data real? Statistical methods for the detection of data fabrication in clinical trials. *British Medical Journal* 331: 267–270.

4. Weir C. & Murray G. (2011). Fraud in clinical trials: detecting it and preventing it. *Significance* 8: 164-168.

5. Pogue JM, Devereaux PJ, Thorlund K & Yusuf S. (2013). Central statistical monitoring: Detecting fraud in clinical trials. *Clinical Trials* 10: 225-235.

6. Godlee F, Smith J & Marcovitch H. (2011). Wakefield's article linking MMR vaccine and autism was fraudulent. *British Medical Journal* 342: 64-66.

7. Bailey KR. (1991). Detecting fabrication of data in a multicenter collaborative animal study. *Controlled Clinical Trials* 12: 741-752.

8. Hixson JR. (1976). *The Patchwork Mouse*. New York: Anchor Press.

9. Baggerly KA & Coombes KR. (2009). Deriving chemosensitivity from cell lines: Forensic bioinformatics and reproducible research in high-throughput biology. *The Annals of Applied Statistics* 3: 1309-1334.

10. Kolata, G. (2011, July 7). How bright promise in cancer testing fell apart. *The New York Times*, p. A1. http://www.nytimes.com/2011/07/08/health/research/08genes.html.

11 Kolata, G. (2011, July 18). Add patience to a leap of faith to discover cancer signatures. *The New York Times*, p. D1. http://www.nytimes.com/2011/07/19/health/19gene.html.

12 Fang FC, Steen RG & Casadevall A. (2012). Misconduct accounts for the majority of retracted scientific publications. *Proceedings from the National Academy of Sciences* 109: 17028–17033.

13 Grieneisen ML & Zhang M. (2012). A comprehensive survey of retracted articles from the scholarly literature. *Public Library of Science (PLoS) ONE* 7(10): 1-15.

14 TransCelerate BioPharma Inc. (2013). Position paper: Risk-based monitoring methodology. Available at: http://transceleratebiopharmainc.com/.

15 US Food & Drug Administration. (2013). Guidance for industry: Oversight of clinical investigations - a risk-based approach to monitoring. Available at: http://www.fda.gov/downloads/Drugs/.../Guidances/UCM269919.pdf.

16 Baigent C, Harrell FE, Buyse M, Emberson JR & Altman DG. (2008). Ensuring trial validity by data quality assurance and diversification of monitoring methods. *Clinical Trials* 5: 49–55.

17 Evans S. (2001). Statistical aspects of the detection of fraud. In: Lock S & Wells F, eds. *Fraud and Misconduct in Biomedical Research, Third Edition*. London: BMJ Books.

18 Venet D, Doffagne E, Burzykowski T, Beckers F, Tellier Y, Genevois-Marlin E, Becker U, Bee V, Wilson V, Legrand C & Buyse M. (2012). A statistical approach to central monitoring of data quality in clinical trials. *Clinical Trials* 9: 705-713.

19 Zink RC, Wolfinger RD & Mann G. (2013). Summarizing the incidence of adverse events using volcano plots and time windows. *Clinical Trials* 10: 398-406.

20 Stokes ME, Davis CS, Koch GG. (2012). *Categorical Data Analysis Using SAS, Third Edition*. Cary, NC: SAS Institute, Inc.

21 Benjamini Y & Hochberg Y. (1995). Controlling the false discovery rate: A practical and powerful approach to multiple testing. *Journal of the Royal Statistical Society Series B: Statistical Methodology* 57: 289–300.

22 Westfall PH, Tobias RD & Wolfinger RD. (2011). *Multiple Comparisons and Multiple Tests Using SAS, Second Edition*. Cary, NC: SAS Institute, Inc.

23 Turner JR & Durham TA. (2009). *Integrated Cardiac Safety: Assessment Methodologies for Noncardiac Drugs in Discovery, Development, and Postmarketing Surveillance*. Hoboken, NJ: John Wiley & Sons.

24 Shah S, Spedding C, Bhojwani R, Kwartz J, Henson D & McLeod D. (2000). Assessment of the diurnal variation in central corneal thickness and intraocular pressure for patients with suspected glaucoma. *Ophthalmology* 107: 1991-1193.

25 Mahalanobis PC. (1936). On the generalised distance in statistics. *Proceedings of the National Institute of Sciences of India* 2: 49–55.

26 Samples and SAS Notes. Sample 30662: Mahalanobis distance: from each observation to the mean, from each observation to a specific observation, between all possible pairs. http://support.sas.com/kb/30/662.html

27 Akhtar-Danesh A & Dehghan-Kooshkghazi M. (2003). How does correlation structure differ between real and fabricated data-sets? *BMC Medical Research Methodology* 3(18): 1-9.

28 Taylor RN, McEntegart DJ & Stillman EC. (2002). Statistical techniques to detect fraud and other data irregularities in clinical questionnaire data. *Drug Information Journal* 36: 115-125.

29 Wu X & Carlsson M. (2011). Detecting data fabrication in clinical trials from cluster analysis perspective. *Pharmaceutical Statistics* 10: 257-264.

30 O'Kelly M. (2004). Using statistical techniques to detect fraud: a test case. *Pharmaceutical Statistics* 3: 237-246.

Detecting Patient Fraud

5.1 Introduction ... 179
5.2 Initials and Birthdate Matching ... 180
5.3 Hierarchical Clustering of Pre-Dosing Covariates Across Clinical Sites ... 185
5.4 Review Builder: Quality and Fraud ... 192
5.5 Final Thoughts ... 197
References ... 198

5.1 Introduction

In Chapter 4, I presented ways to identify various data anomalies that are specific to the clinical site: how the site adheres to the trial protocol, how site personnel collect and report data, and even how equipment may be performing. In this chapter, we focus on the study participants. Patient behavior can greatly affect the outcome of a clinical trial. This is true if patients are noncompliant in taking their study medication, or if they fail to properly complete or maintain a study diary to record various symptoms or episodes of their disease (a familiar example is the patient who completes his or her study diary in the clinic parking lot just prior to a visit). In this chapter, we focus solely on identifying patients who enroll at two or more clinical sites. Enrolling at more than one trial site is a particularly troublesome form of misconduct; these patients may be seeking additional access to a study drug, additional financial compensation for participating in the trial, or continued access to high-quality health care.

Multiple enrollments raise a number of concerns for the study team. Such behavior can contribute to a severe safety event for the patient, particularly if the subject is enrolled in two or more sites in parallel or randomized to multiple active treatments (assuming they are compliant with their treatment in the first place). For statisticians, it violates the statistical independence often assumed between trial participants for the analysis of trial data. In practice, however, such events are likely rare enough that any impact to analysis assumptions should be minimal. The practical implication of these multiple enrollments is that it creates an accounting and reporting nightmare for the study team, especially if the subject happens to be randomized to different treatments. Sensitivity analyses removing these subjects may be expected by regulatory agencies for the final clinical report; the resulting loss of the sample could affect the power for treatment comparisons (though any difference in the interpretation of results with or without these patients may be problematic). Ideally, these instances should be identified as early as possible to minimize the amount of data affected, avoid potential toxicity, and allow for the enrollment of additional subjects to meet sample size requirements.

While the above discussion focuses on multiple enrollments within an individual trial, the same concerns are often relevant for the clinical development program. Later phase trials often have criteria that exclude individuals from participating in a second study, either through direct means (e.g., subjects who have previously participated in a trial for drug X) or indirect means (e.g., subjects must be treatment naïve to participate). Multiple enrollments across two or more trials generate extra work for the study team to appropriately summarize drug exposure and safety for these individuals for integrated summary reports or meta-analyses.

So why make an effort to identify these subjects at all? Some may take the attitude that ignorance is bliss - in that identifying these subjects, or any quality issues for that matter, only creates work and identifies problems that regulators can point to later during review. This is an unfortunate point of view. Protecting the well-being of the trial participants is of the utmost importance, as is preserving trial integrity. If the study team is proactive and regularly assesses the quality of trial data, problems can be identified early and resolved while the trial is ongoing. This is far preferable to being surprised by a quality issue after database lock, particularly if it is brought to the attention of the sponsor by a regulatory agency. In short, simply ignoring the problem does not make it go away.

Given that patients enrolled in the **Nicardipine** trial as a result of a subarachnoid hemorrhage (SAH), it is extremely unlikely they would have had the opportunity to enroll at multiple clinical sites. Despite the perceived limitation of this trial for identifying subjects with multiple enrollments, we can still use the data to motivate the discussion, describe the methodologies, and illustrate the functionality of JMP Clinical for identifying the same patient at multiple clinical sites. In "5.2 Initials and Birthdate Matching" on page 180, I present methods to match subjects using their birthdates or initials. In the absence of such data, "5.3 Hierarchical Clustering of Pre-Dosing Covariates Across Clinical Sites" on page 185 describes an analysis to calculate the similarity between pairs of subjects within subgroups defined by important demographic characteristics. "5.4 Review Builder: Quality and Fraud" on page 192 illustrates how to create a data quality and fraud review using the **Review Builder** that can be generated each time the study database is updated.

5.2 Initials and Birthdate Matching

Perhaps the most straightforward means of identifying subjects who have enrolled multiple times in a clinical trial is to match patients with similar birthdates or initials. At least, this is a very useful first step. The Birthday Problem, known to any student who has taken a course in probability, states that among 50 people there is a 97 percent chance that at least two individuals share the same birthday [1]. In fact, there is a 50-50 chance of a match with only 23 subjects in the clinical trial. This result is based on the assumption that each and every birthday is equally likely, which is not true in practice. The Birthday Problem does illustrate, however, that birthday matches are not as rare as one might think.

Despite the potential to match distinct individuals based on birthdate (or initials) alone, the demographic and physical characteristics of these patients can be used to eliminate pairs that are unlikely to be the same person. The **Birthday and Initials** report identifies matches based on birthdate (DM.BRTHDTC) or initials (records from the subject characteristics or SC domain where SC.SCTESTCD = "SUBJINIT"). Users who select birthdate as matching criteria can opt to allow for a window around the birthdate (in case dates were entered into the database incorrectly). Matching by initials allows for the possibility of a missing middle initial (in other words, RCZ and R-Z would be considered a match).

Figure 5.1 List of Subjects with Duplicates for Birthday

Matching Variable Value	Unique Subject Identifier	Study Site Identifier	Date/Time of Birth	Sex	Race	N
15Apr1915	141001	14	1915-04-15	M	WHITE	1
	442001	44	1915-04-15	F	WHITE	1
01Jan1917	392012	39	1917-01-01	F	OTHER	1
	41018	04	1917-01-01	M	WHITE	1
11Aug1925	221020	22	1925-08-11	M	BLACK OR AFRICAN AMERICAN	1
	282003	28	1925-08-11	F	WHITE	1
28Jul1928	241003	24	1928-07-28	F	WHITE	1
	81011	08	1928-07-28	F	WHITE	1
13Nov1935	141021	14	1935-11-13	F	OTHER	1
	231027	23	1935-11-13	F	WHITE	1
25Aug1937	31006	03	1937-08-25	M	WHITE	1
	31012	03	1937-08-25	F	WHITE	1
23Mar1938	171001	17	1938-03-23	M	WHITE	1
	271014	27	1938-03-23	F	WHITE	1
29May1939	11034	01	1939-05-29	F	WHITE	1
	161003	16	1939-05-29	F	WHITE	1
21Jan1941	282008	28	1941-01-21	F	BLACK OR AFRICAN AMERICAN	1
	351006	35	1941-01-21	M	WHITE	1
09Jul1941	11033	01	1941-07-09	F	WHITE	1
	392017	39	1941-07-09	F	WHITE	1
27Aug1943	11048	01	1943-08-27	F	WHITE	1
	291002	29	1943-08-27	F	ASIAN	1
22Feb1947	201019	20	1947-02-22	M	WHITE	1
	81023	08	1947-02-22	M	WHITE	1
23Jan1949	141013	14	1949-01-23	F	ASIAN	1
	461023	46	1949-01-23	M	WHITE	1
09Feb1954	161016	16	1954-02-09	F	WHITE	1
	241011	24	1954-02-09	F	WHITE	1
15Oct1955	141054	14	1955-10-15	F	WHITE	1
	231020	23	1955-10-15	F	WHITE	1
18Oct1956	91038	09	1956-10-18	F	BLACK OR AFRICAN AMERICAN	1
	91040	09	1956-10-18	F	BLACK OR AFRICAN AMERICAN	1

Based on the default options of **Birthdays and Initials**, Figure 5.1 on page 181 lists 16 birthdates shared by at least two patients from the **Nicardipine** study, including demographic information for gender (DM.SEX) and race (DM.RACE). If available, data for ethnicity (DM.ETHNIC) or physical characteristics such as height, weight, or body mass index would be reported as well (records where VS.VSTESTCD = "HEIGHT", "WEIGHT" or "BMI"). Here, differing sex and race within a match can quickly identify pairs that do not require further attention. For example, the first two rows representing 15Apr1915 are likely not the same subject since the gender is different. However, the pair for date 28Jul1928 (rows 7 and 8) matches in both race and gender and may require further review. Notice the pair of subjects in the last two rows for 18Oct1956. They match on race and gender and are participating at the same clinical site! The **Show Subjects** drill down allows the user to review additional data for these patients from the DM or ADSL data set, including the date when these subjects enrolled in the trial.

How do we explain this last pair of subjects (91038 and 91040)? Perhaps the subject was initially a screen failure under ID 91038 and was successfully enrolled under 91040. For example, maybe the subject had to wash out of medication that initially excluded them from participating in the trial. This is unlikely; the analysis population defaults to the Safety Population, implying that both subject IDs received study drug. Another possibility is that twins enrolled at this site; such an occurrence may even come with a match for their initials (if collected)! This is unlikely for **Nicardipine** given the SAH entry criteria for subjects in this trial, but an instance of twins enrolling at the same site is certainly plausible under differing circumstances.

One potentially troubling case is a center that enrolls a subject twice for the financial benefit of treating an additional subject. Another possibility, problematic but with good intentions, may be due to what some consider compassionate use. In the absence of an effective therapy for the disease under investigation, or for subjects who have failed available therapies, the investigator may enroll a patient a second time to give him a chance at continued treatment. Of course, this assumes that the trial sponsor would not have provided some form of access to the treatment once patients completed the clinical trial (regardless of their randomized treatment). Whatever the reason, such a pair would warrant additional investigation. To generate a review more informative than what is available using the **Show Subjects** drill down, select the two subjects using the Matching Variable Value slider from the **Data Filter**. This action subsets the table to those individuals who meet selected criteria (Figure 5.2 on page 182). Now click **Profile Subjects**.

Figure 5.2   Birthdate Match Within a Clinical Site

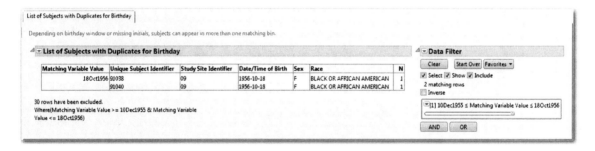

Initially, a patient profile is opened for subject 91038 only. However, holding the Ctrl key while selecting 91040 from the **Subjects** list box will summarize both patients in the profile simultaneously (Figure 5.3 on page 183) after hiding the results of several domains, as described below. This provides the analyst with a graphical means to compare subject data side by side. Particularly informative data may include concomitant medications and medical history; these individuals have some overlap for these domains, though the reported terms may be common for individuals with an SAH. As an alternative to the graphical presentation of the profile, go to the **Tables** tab (Figure 5.4 on page 184). The data for these two patients is summarized by domain. Unlike static tables, the user can right-click on the variables within each table and go to **Sort by Column** to sort rows so that similar records between subjects are side by side for easier review. The user may also drop variables from the tables by right-clicking and selecting **Columns**, or by unchecking the appropriate variables under **Data Tables** on the left-hand side. Dropping columns may be useful when printing or sharing PDF or RTF versions of these tables using the **Share Report** button, to prevent wrapping of extra-wide tables.

Any number of subjects may be selected and viewed simultaneously in the patient profile, though screen real estate will be the limiting factor for the number of subjects that can be reviewed at one time. For displaying multiple subjects, the user can limit the data summarized in the graphical profile by going to the **Domains** outline box and selecting the particular domains or records of interest; similar results are possible for the **Tables** tab by going to the **Data Tables** outline box. Users may save their patient profile configurations for later use within this or other studies by going to **Templates > Create or Update a Template**.

Figure 5.3   Patient Profile Simultaneously Displaying Two Individuals

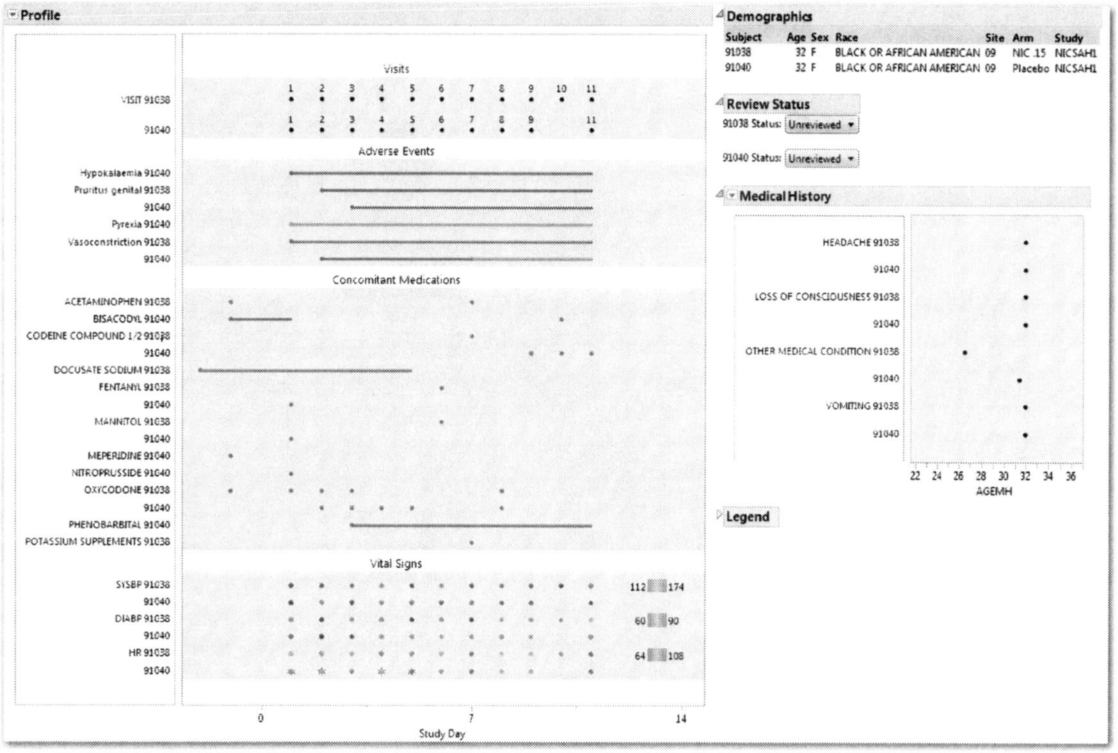

This analysis can also be used to identify subjects across multiple clinical trials. However, this currently requires your programming team to combine multiple studies into a single "study" that can then be registered within JMP Clinical. This combination study should contain the DM and VS domains of all trials to supply the necessary demography and physical characteristics. If initials are available, the relevant rows of the SC domain should be included. Additional data for medical history and concomitant medications can be included to provide more informative patient profiles if any matches appear particularly interesting. In lieu of asking the study programmers to combine data, any user can combine the data from trials using **Studies > Combine Studies** (Figure 5.5 on page 185). This feature is most useful for studies with very similar protocols, as the individual domains are "stacked" to create the new combined domains. Users are encouraged to exercise

184 Chapter 5 / Detecting Patient Fraud

caution, since data will be summarized assuming consistency of units and coding between the combined studies. The limitations of this analysis are that the individual studies must already be registered within JMP Clinical, and only two studies may be combined at any one time. Users specify the combined study name as well as the directories in which to store the combined data sets.

Figure 5.4 Tabular Patient Profile Displaying Two Individuals

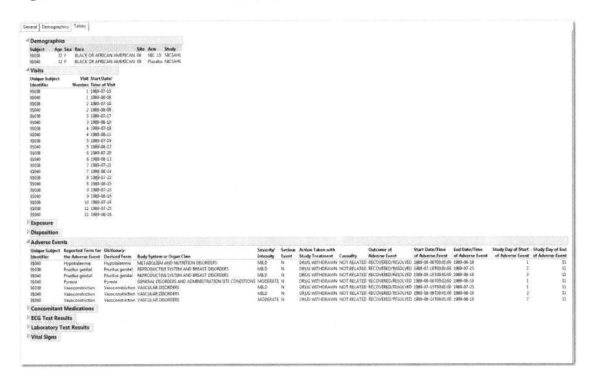

Figure 5.5 Combine Study Dialog

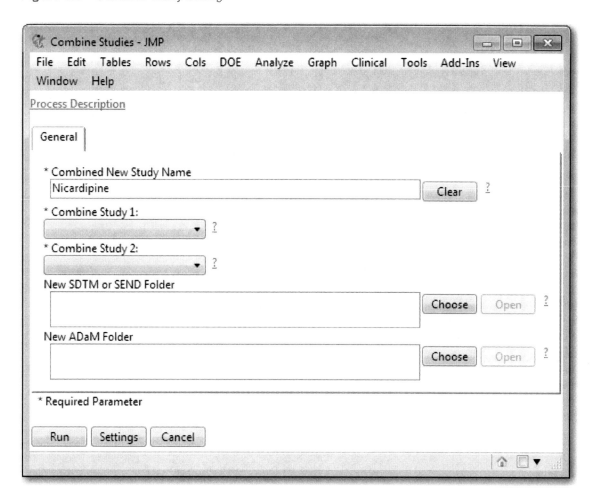

5.3 Hierarchical Clustering of Pre-Dosing Covariates Across Clinical Sites

The analyses of the previous section are relatively straightforward, but rely on the availability of very specific data about the trial participants. Often, birthdates and initials may not be present in the study database over concerns that providing such information violates patient privacy. In the absence of such data, we need alternate methods to identify multiple enrollments. But fear not! There is a wealth of data that can be used to identify similar subjects, though the best data to use may depend on the individuals under study. For example, height is a very useful measurement for adults that should remain relatively constant over time (especially if clinical sites are diligent about collecting this data after the removal of patients' shoes). Weight is a possibility, but can naturally vary between 2 to 4 pounds a day depending on diet and exercise patterns, and fluctuate more significantly due to lifestyle changes, the underlying disease, or the effects of the current therapy [3]. However, such straightforward data may not be collected if it has little to do with the disease or its treatment. Medications used for the treatment of chronic disease are potential candidates as long as they are not excluded by the protocol. The most useful medications would likely need to be for a disease relatively distinct from the one under study; otherwise, many trial participants would be using them to address various symptoms. Matches based on past surgeries or other procedures are possible, provided the medical history was reported accurately in the first place. Finally, there are those data that are collected from the various tests and procedures performed to measure the safety and efficacy of the experimental treatment within the clinical trial.

In Section 4.4.2, I described an analysis to identify fabricated patients. The analysis is most useful when the fabricated patient is a modified copy of another participant within the clinical site, since a distance calculation between the covariates for this pair of subjects will identify them as being overly similar when compared to other pairs of subjects. In this section, we take a comparable approach to calculate the similarity between pairs of subjects within the trial in order to identify multiple enrollments. This can lead to a lot of comparisons! The number of comparisons was limited in Section 4.4.2, since they were performed only between patients within the same clinical site. Using **Nicardipine** as an example, examining all pairs of subjects creates (906 x 905) / 2 = 409,965 pairwise distances. This is a great deal of information to sift through and interpret. We can restrict the number of comparisons by limiting pairs within demographic subgroups based on gender, race, and country (DM.SEX, DM.RACE and DM.COUNTRY, respectively). Another difference from the previous analysis involves the data used in computing the distance metric. Subject to dialog selections, all of the data was used for detecting fabricated patients. Here, the data is confined to those measurements collected prior to the first dose of study drug (active or placebo) to avoid the potential effects any trial therapy may have on the outcomes of various procedures.

Figure 5.6 Cluster Subjects Across Study Sites Dialog

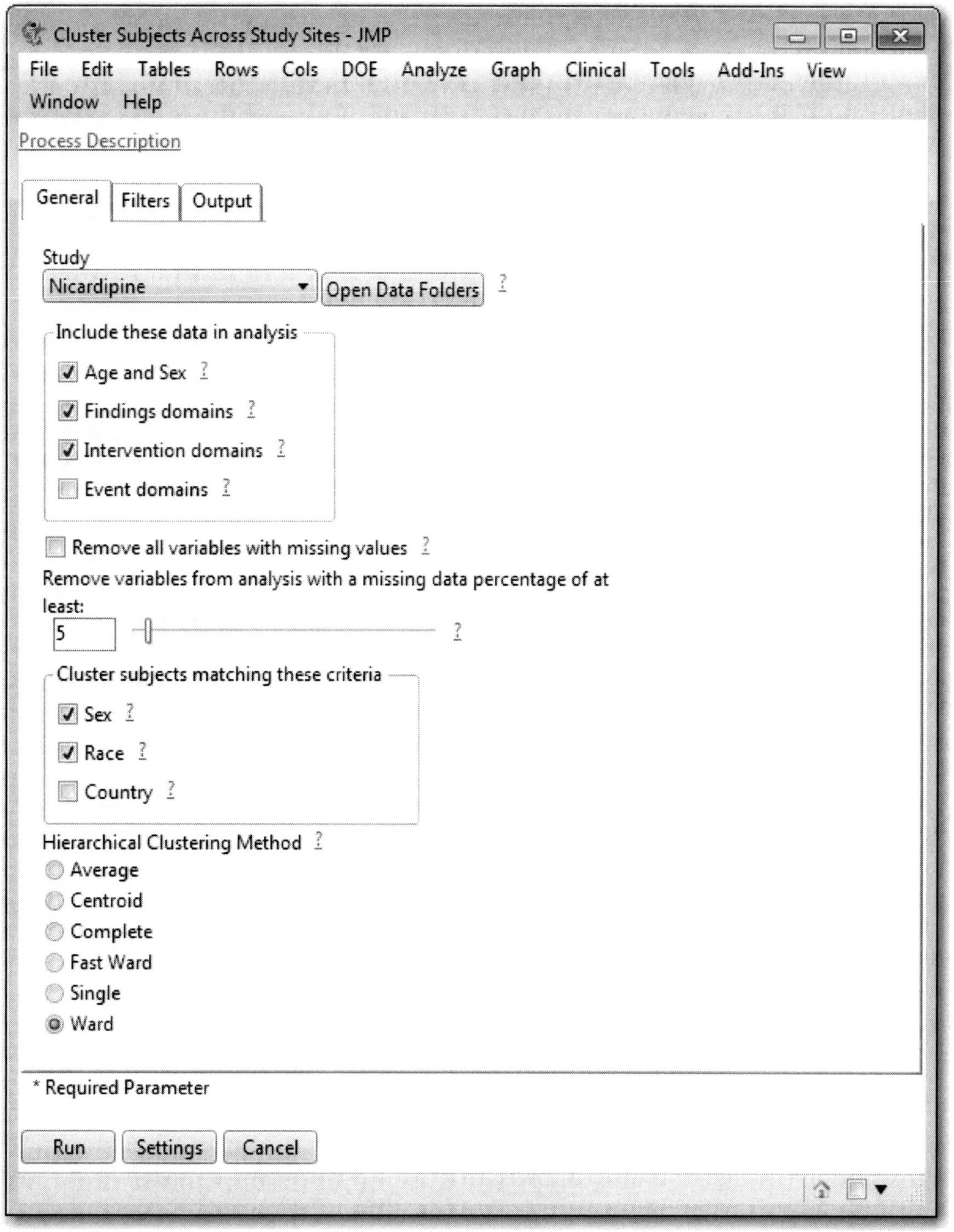

Figure 5.6 on page 187 presents the dialog for the **Cluster Subjects Across Study Sites** report. Available options are similar to those for **Cluster Subjects Within Study Sites**. The user may choose to **Include these data in the analysis**:

- Age and Sex can be culled from the DM distribution (DM.AGE and DM.SEX, respectively);
- Variables for lab tests, vital signs, symptoms, and other Findings data will be generated by visit number and planed time point number (xx.VISITNUM and xx.xxTPTNUM) and included in the analysis if the Findings domains is checked (using xx.xxSTRESN, the Numeric Result/Finding in Standard Units);
- Variables representing frequencies of medications, tobacco use, and other interventions domains will be included by selecting Interventions domains (xx.xxDECOD for Standardized Names);
- Variables representing Numeric Result/Finding in Standard Units frequencies of adverse events, medical history terms, and other event domains will be included by selecting Event domains (xx.xxDECOD for Standardized Names).

Because we calculate Euclidian distance directly on the covariates, the analysis can accommodate missing values (distances are calculated where covariates are non-missing for both subjects in a pair). By default, however, the dialog excludes variables where the percentage of missing values exceeds five percent across the population of subjects determined by the **Filters** tab. Alternatively, any variables with missing data can be removed from the analysis. Users can select a **Hierarchical Clustering Method** (default Ward). For details on the various clustering methods, users are encouraged to review the technical details within the Hierarchical Clustering description within **Help > Help Contents**. Finally, the option **Cluster subjects matching these criteria** limits the comparisons of subjects to within demographic subgroups; sex and race are selected by default. **Run** the analysis based on the selections in Figure 5.6 on page 187 (select Interventions so that concomitant medications are added to the analysis) to generate the output in Figure 5.7 on page 189.

5.3 Hierarchical Clustering of Pre-Dosing Covariates Across Clinical Sites

Figure 5.7 Between-Subject Distance Summary

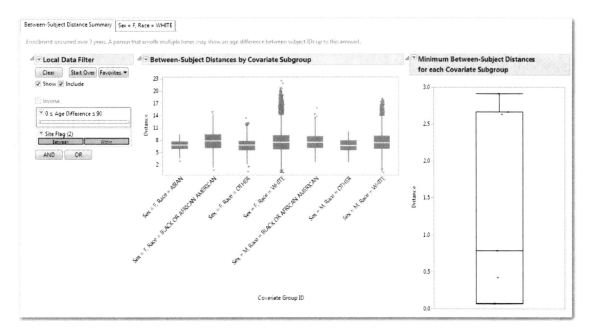

The presentation in Figure 5.7 on page 189 mirrors that of **Cluster Subjects Within Study Sites** very closely, save for two important differences. First, rather than present box plots of pairwise distances by clinical site, the box plots are presented by covariate subgroups, here based on gender and race. The second difference is the addition of a **Local Data Filter** to subset to pairs of subjects with very similar demographic characteristics. Only age (DM.AGE) is available for Nicardipine, but height and weight would be available in the filter if these data were collected within the clinical trial (records in VS where VS.VSTESTCD = "HEIGHT" or "WEIGHT"). If the cluster analysis is not of interest or a bit too difficult to comprehend, the **Local Data Filter** on either summary tab makes it possible to identify patient IDs within a few years, kilograms or centimeters of one another matched according to their gender, race, and country (if selected). Note that enrollment in **Nicardipine** took place from 1987 to 1989, making it possible for a subject's age to have changed by three years over the course of the trial. The **Between** and **Within** values in the filter describe pairs of patients from different sites or the same site, respectively. Clicking **View Data** will open a data table of all possible pairs of subjects with their pairwise distance, covariate subgroup membership, and age, height, and weight differences, if available.

Figure 5.8 Cluster Analysis for White Females

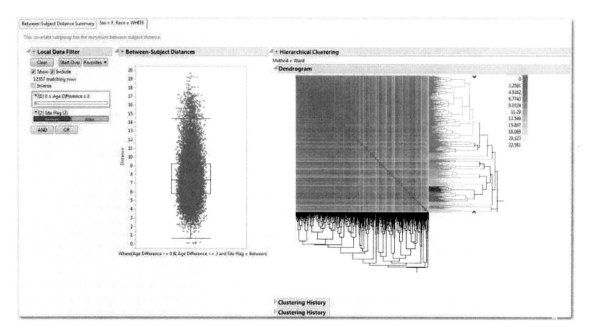

We would not expect to see major differences among the distributions of patient pairs within the covariate subgroups; small sample sizes would likely be the biggest culprit for noticeable differences. However, this presentation is provided for a reason similar to that in Section 4.4.2: to provide the analyst with a frame of reference of how similar is too similar between any pair of subjects. The box plot on the right side of the **Between-Subject Distance Summary** tab summarizes the individual pair from each covariate subgroup with the minimum distance between them. Dragging the cursor to select the minimum point within this box plot selects two overlapping points; clicking **Show Subjects** opens a data table of white females and males. The cluster analysis Tab for the first of these groups (here, white females) sharing this minimum value is opened for display. Other groups can be opened by going to the **Tabs** outline box, selecting a covariate subgroup, and clicking **View Tab**.

The **Sex = F, Race = WHITE** tab presents two figures (Figure 5.8 on page 190). The one at left is a box plot of all pairwise distances for white females. The one at right contains the hierarchical clustering using the Ward method for combining clusters and a heat map to graphically describe the magnitude of the pairwise distances, where the white to red legend indicates less to greater similarity, respectively. The **Show Subjects** drill down will open a data table of the subjects selected from the box plot. Use the **Local Data Filter** to limit pairs of subjects based on an age difference of three years, and click **Between** to subset to pairs where the subjects were from different clinical sites. Drag and select the most similar pairs of subjects and click **Show Subjects**. You will notice from the data table that many of the subjects are missing systolic and diastolic blood pressures (Figure 5.9 on page 191). This means that their distance was calculated solely based upon their ages and frequency of medications, which may help explain the perfect matches between pairs of subjects (i.e., distances of zero).

5.3 Hierarchical Clustering of Pre-Dosing Covariates Across Clinical Sites

Figure 5.9 White Females with Small Pairwise Distances

	Unique Subject Identifier	Study Site Identifier	Age	Race	Description of Planned Arm	Country	Covariate Group ID	DIABP_V1	SYSBP_V1	ACETAMINOPHEN	AMINDOAPROIC ACID	BENZOE
1	141003	14	73	WHITE	NIC .15	USA	Sex = F, Race = WHITE	70	162	0	0	
2	201008	20	57	WHITE	NIC .15	USA	Sex = F, Race = WHITE	66	118	0	0	
3	21004	02	70	WHITE	NIC .15	USA	Sex = F, Race = WHITE	.	.	0	0	
4	231027	23	53	WHITE	Placebo	USA	Sex = F, Race = WHITE	78	135	0	0	
5	271008	27	71	WHITE	NIC .15	USA	Sex = F, Race = WHITE	.	.	0	0	
6	392013	39	55	WHITE	Placebo	USA	Sex = F, Race = WHITE	.	.	0	0	
7	41010	04	76	WHITE	Placebo	USA	Sex = F, Race = WHITE	.	.	0	0	

Alternatively, we can select the most similar pair or pairs of subjects (those closest to zero) from the box plot and highlight their rows in the heat map by clicking the **Show Rows in Heat Map** drill down. This can be useful for identifying the cluster membership for any selected pairs of subjects within the **Hierarchical Clustering** analysis, or for identifying groups of 3 or more that could indicate that the subject has enrolled more than twice. Unlike the analysis within centers from Section 4.4.2, it may be extremely difficult to view row and column labels of the heat map and dendrogram based on the number of patients within the subgroup. There are view tools available, however, to help focus in on areas of interest. For example, selecting lines in the dendrogram will highlight corresponding clusters (Figure 5.10 on page 191); go to the red triangle menu and click **Zoom to Selected Rows** to enlarge the area (Figure 5.11 on page 192). **Release Zoom** under the red triangle will return the figure to normal. Alternatively, select the **Magnifier** (Figure 5.12 on page 192) in the JMP Tools Icon Bar, hover over the part of the dendrogram you wish to enlarge, and click the mouse button. Repeated clicks may be necessary to zoom to an appropriate level. To release the zoom, hold the Alt key and click once on the dendrogram. The **Magnifier** magnifying glass is also useful to zoom into areas of the box plot where subjects may be more similar. Do not forget to re-select the **Arrow** tool from the Tools icon bar once you are finished zooming. Finally, **Subset Clustering** can recalculate the cluster analysis for the subset of heat map rows (subjects) selected manually or through the **Show Rows in Heat Map** drill down; **Revert Clustering** returns the analysis to its original state.

Figure 5.10 Selecting a Cluster for Zooming

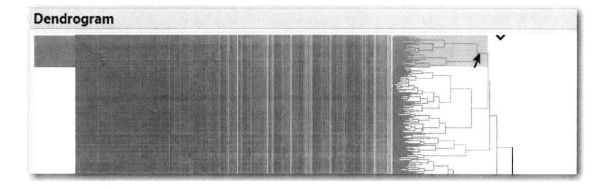

192 Chapter 5 / Detecting Patient Fraud

Figure 5.11 Using Zoom to Selected Rows

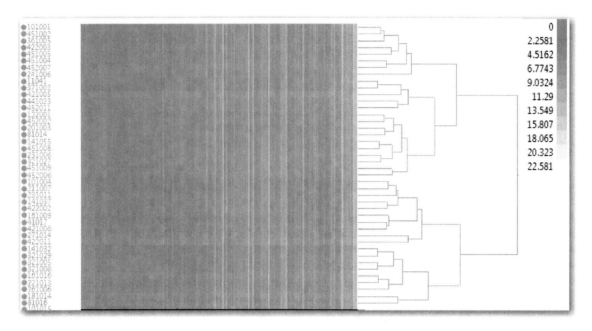

As described in the previous section, this analysis can also be used to identify subjects who have enrolled in multiple clinical trials. Again, this currently requires your programming team to combine multiple studies into a single "study" that can then be registered within JMP Clinical. Based on the number of trials and the total number of subjects, it may be worthwhile to limit the data to demographic and physical characteristics (DM and rows of the VS domain, respectively), and to any "hard" measurements collected at the site during clinic visits that were available prior to dosing with study drug.

Figure 5.12 Magnifier Icon

5.4 Review Builder: Quality and Fraud

In this section, I summarize a feature that can be used by the reports in this and the previous chapter. In fact, **Studies > Review Builder** can incorporate most of the reports available within

JMP Clinical. Throughout the book, you may have noticed how each analysis performed typically involves opening a JMP Clinical report dialog, selecting a set of options, and then running the report to review and explore the output. While this is a straightforward way to conduct analysis, it can be become quite tedious to repeat the same actions again and again over time for the same clinical trial. Within the pharmaceutical industry, this phenomenon tends to happen because of the manner in which data are collected. Data are collected for a time, and a snapshot of the data is taken for the team to conduct a review. Queries are sent to the clinical sites to address any problems or inconsistencies with the data. When sufficient new data are collected, the process is repeated: a snapshot is taken, new data are reviewed, and modifications to previous data are examined for correctness. Naturally, we may want to regenerate a pre-specified set of analyses each time the data is reviewed. This is where the **Review Builder** comes into play. I discuss data snapshots more in Chapter 6.

In Section 3.4, I showed how to build some straightforward JMP Add-Ins to automate a set of repeated actions for risk-based monitoring (RBM). Unlike a JMP Add-In, the **Review Builder** allows the analyst to bundle together a set of JMP Clinical reports with the appropriate set of options selected so that this set of reports (the review) can be run en masse when the clinical trial data has been updated. Such reviews are beneficial for individuals who are not as comfortable selecting all of the options necessary to conduct the analysis appropriately. Reviews can be built by someone knowledgeable and comfortable constructing the analysis and shared with the larger team. Click **Studies > Review Builder**. The **Clinical Review Builder** window is very similar to the JMP Clinical Starter, with the addition of a window for selecting reports for analysis (Figure 5.13 on page 194). JMP Clinical ships with three reports: Medical Monitoring, Quality and Fraud, and Signal Detection, which is a set of analyses for comparing adverse events between treatment arms. Under **Current Review**, select the **Quality and Fraud Review** and click **Run**. This review contains the 9 **Fraud Detection** reports currently available with JMP Clinical, which are run in sequence for the **Current Study**.

Figure 5.13 Clinical Review Builder

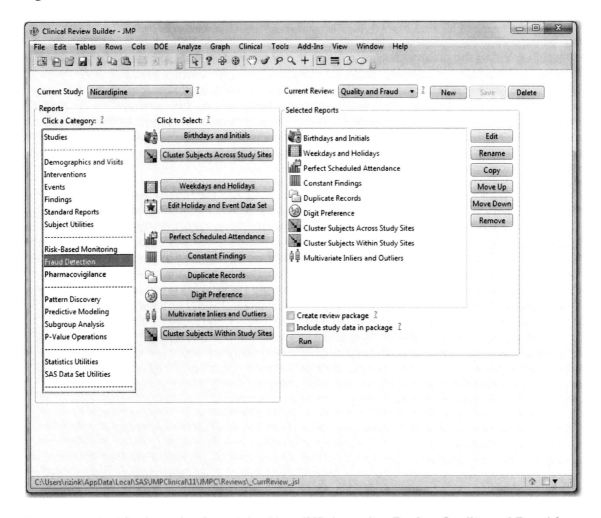

Summary output for the review is contained in a JMP Journal as **Review Quality and Fraud for Study Nicardipine** (Figure 5.14 on page 195). Each button in the Journal corresponds to the output of a single JMP Clinical report. For example, clicking on **Birthdays and Initials** will open the tabbed report presented in Figure 5.1 on page 181. This or any review can be applied to any study registered within JMP Clinical. However, if certain data are not available based upon selected reports or requested options within those reports, error messages will appear. In this case, create a new report altering the contents of an existing report to be appropriate for the study of interest. For example, click **New** to create a report and provide a **New Review Name** (e.g., Patient Fraud); clicking **OK** with Quality and Fraud highlighted under **Existing Reviews** and **Add reports from the reviews selected above** selected will add all reports from the Quality and Fraud review to the Patient Fraud review.

Figure 5.14 Quality and Fraud Review for Nicardipine

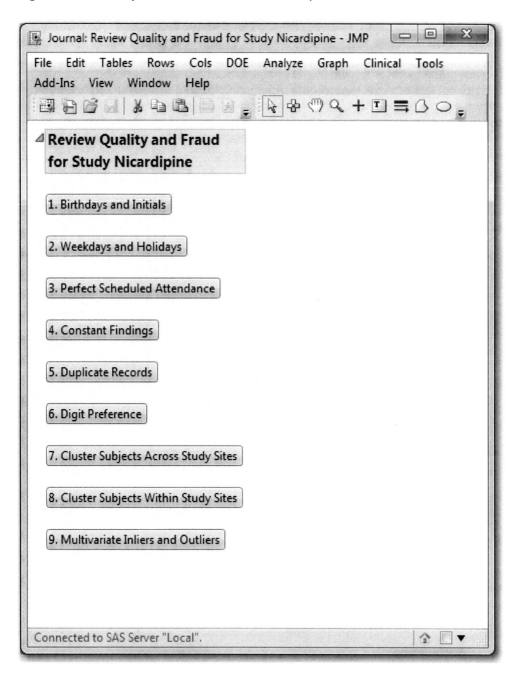

Under the Patient Fraud review, **Remove all** reports save for **Birthdays and Initials** and **Cluster Subjects Across Study Sites**. **Add** a second instance of **Birthdays and Initials** by going to the **Fraud Detection** menu in the Clinical Starter and clicking **Birthdays and Initials**. For this instance of the report, click **Edit** to modify the options to match patients based on their initials.

Save by clicking **Update Review**. Click **Move Up** so that this analysis occurs prior to the **Cluster Subjects Across Study Sites**. **Rename** both of these analyses so that their titles are more informative (Figure 5.15 on page 196). Further, click **Edit** for **Cluster Subjects Across Study Sites** to add **Intervention domains** to the analysis. Save by clicking **Update Review**. Click **Save** to save the review, then click **Run**. Based on our discussions earlier in this chapter, it should come as no surprise that an error is thrown by the analysis matching subjects based on initials, since these data are not available for **Nicardipine**. Attempting to open the output for this report from the journal will re-open the ERROR window that occurred when the analysis was trying to generate output (Figure 5.16 on page 197). While this example built a new review from an existing one, I could have easily started from scratch by deselecting **Add reports from the reviews selected above** when creating the Patient Fraud review.

Figure 5.15 New Patient Fraud Review

Figure 5.16 Patient Fraud Review Displaying Error for Initial Matching

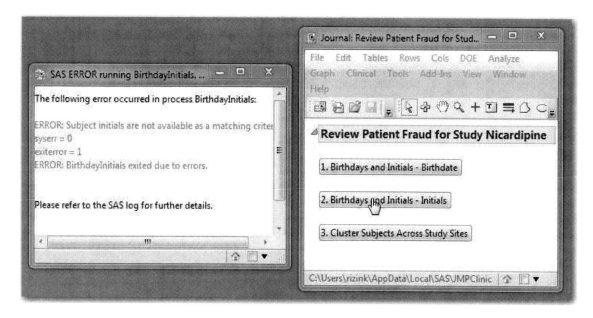

The last feature I want to highlight in this section involves the option **Create review package**. When this option is selected and a review is generated, a JMP Add-In is created and placed in the output folder of the **Current Study**. This Add-In can be shared with other users through email or by being placed in a directory that others can access. Double-clicking on the file adds the review journal to the **Add-In** menu, and the reports in this analysis can be viewed at any time. When operating in local mode, you may want to **Add study data in package** so that users receiving the Add-In have the ability to use drill downs that require the study database, such as **Patient Profile**. In this scenario, the Add-In will first register the study using **Add Study from Folders**. When operating in server mode, users will have access to all drill down functionality from the review since the registered study is available to everyone with access. This Add-In can be removed from the menu by going to **View > Add-Ins** and deselecting **Enabled** to temporarily turn it off, or clicking **Unregister** to remove it permanently. In Chapter 6, I'll talk more about updating study data in JMP Clinical when a new snapshot becomes available.

5.5 Final Thoughts

In this chapter, we focused on identifying patients that enrolled at two or more clinical sites within the same study, and described how the analysis could be extended to identify those individuals that enrolled within multiple studies within the same clinical development program. These analyses suggest the usefulness of collecting height and weight for all trials, not just those studies where such physical characteristics affect the dose of study drug or play an important role in the disease. There are numerous other checks that could be performed to assess the quality of data provided by study participants. Currently, the **Constant Findings** report summarizes any tests within Findings domains with no variability, while the **Duplicate Records** analysis summarizes repeated sets of measurement values within and across subjects. Such analyses may be useful to identify any unusual trends within patients as to how they are completing questionnaires or diaries. Using **Analyze > Multivariate Methods > Multivariate and Correlations** on patient-provided data from the data set (open with **View Data**) generated from either **Multivariate Inliers and Outliers** or **Findings Time Trends** may identify any subjects with unusual patterns of reporting.

Of course, there is more to be done in this area to assess the compliance of patients and quality of data they report. For example, future development examining the correlation structure of related measurements may be used to identify subjects (and even sites) exhibiting unusual relationships among covariates [4,5]. Further, identifying patients who have interruptions in their dosing early on can provide the investigator with the opportunity to counsel the patient to prevent longer or repeated dosing interruptions. The benefit of using CDISC standards is that review tools can be available earlier in the life of the study, giving the team the chance to intervene and apply a course correction, rather than summarize the extent of the problem at the end of the trial.

References

1. Ross S. (2014). *Introduction to Probability Models, 11th Edition*. Oxford, UK: Academic Press.

2. Buyse M, George SL, Evans S, Geller NL, Ranstam J, Scherrer B, LeSaffre E, Murray G, Elder L, Hutton J, Colton T, Lachenbruch P & Verma BL. (1999). The role of biostatistics in the prevention, detection and treatment of fraud in clinical trials. *Statistics in Medicine* 18: 3435-3451.

3. Bauer, H. (2012, August 29). Beware the Scale: Learn the Right Way to Weigh. *Eat+Run Blog*. http://health.usnews.com/health-news/blogs/eat-run/2012/08/29/beware-the-scale-learn-the-right-way-to-weigh

4. Akhtar-Danesh A & Dehghan-Kooshkghazi M. (2003). How does correlation structure differ between real and fabricated data-sets? *BMC Medical Research Methodology* 3(18): 1-9.

5. Taylor RN, McEntegart DJ & Stillman EC. (2002). Statistical techniques to detect fraud and other data irregularities in clinical questionnaire data. *Drug Information Journal* 36: 115-125.

6

Snapshot Comparisons

6.1 Introduction	199
6.2 Domain Keys	200
6.3 Review Flags	212
6.3.1 Record-Level	212
6.3.2 Patient-Level	214
6.4 Adding and Viewing Notes	217
6.5 Using Review Flags	220
6.5.1 The Domain Viewer	220
6.5.2 Demographic Distribution	225
6.5.3 Patient Profiles	228
6.5.4 AE Distribution	232
6.6 Final Thoughts	237
References	238

6.1 Introduction

Data from a clinical trial should be examined by as many eyes as possible. Each individual brings a unique skill set important for understanding patient safety, protocol adherence or data insufficiencies that can affect the final analysis. Clinical data review is extremely time-consuming; a major reason for this prolonged effort involves the data collection process. To perform the final analysis as early as possible after the clinical trial ends, data are collected and cleaned as they become available. This review may necessitate queries to the clinical site to correct erroneous values or request additional data if the currently available information is unclear or insufficient. So not only are new data reviewed, the responses to queries on previously-available data are also assessed to examine whether or not they sufficiently address the perceived inconsistencies. Further, the totality of the data may need to be scrutinized for consistency in results across time. Tracking the status of any given data point can be useful to minimize the review efforts of the study team. While database management systems (DBMS) can easily manage and identify changes in data over time, these tools are limited to few individuals, and such review features are rarely available for the submission-ready data sets required for analysis. Further, tables and listings supplied to support clinical data reviews are often slightly-modified versions of those analyses that will be used for the final clinical study report. Rarely are resources available to perform additional programming to highlight new or modified values within the output to streamline reviews.

As discussed in Chapter 1, it is not practical to update needed data sets and regenerate review reports daily—it would be difficult for reviewers to cope with a constantly evolving database! Instead, an intermittent "snapshot" of the study database is taken which reflects the currently collected data and any changes since the previous snapshot. The snapshot is reviewed and necessary queries are generated to address any inconsistencies in the data or gaps in the information provided. The frequencies of these snapshots are dictated by enrollment, the study design, patient population, and the time remaining in the clinical trial. Knowing which data have changed or have been received since the previous snapshot not only accelerates the review process, but allows the study team the ability to address any recent issues that require intervention.

In this chapter, I detail the snapshot comparison feature for JMP Clinical with the goal of streamlining the centralized review of your clinical trial. Comparisons between current and previous data snapshots identify new or modified values so that reviewers do not spend excessive time on previously examined data. In "6.2 Domain Keys" on page 200 I describe how such functionality is possible using domain keys. "6.3 Review Flags" on page 212 documents the review flags that are created at the subject and record level. These review flags can be used to easily identify and subset to patients with new data, or pinpoint the records and variables that have experienced one or more modifications during the course of the clinical trial. System- and user-generated notes will be summarized within "6.4 Adding and Viewing Notes" on page 217. I illustrate how the above review features can be used in practice in "6.5 Using Review Flags" on page 220. At this point, interested users may also want to examine the **Review Builder** described in "5.4 Review Builder: Quality and Fraud" on page 192 to easily and repeatedly perform analyses from one snapshot to the next. In the below discussion when I use the term "record," I refer to the rows of any data set. Note that the current implementation of the snapshot comparison feature does not consider data from the Trial Design domains or Relationship datasets (i.e. SUPPxx domains or RELREC).

6.2 Domain Keys

In order to illustrate the snapshot comparisons functionality, it is necessary to add and update a study within JMP Clinical. This task is straightforward when operating in local mode since dialog settings are readily available to access the appropriate data. When operating in server mode, the location of both the **Nicardipine** and **Nicardipine Early Snapshot** (the study to be added) data directories are required (and both sets of data need to be copied to the server). As a reminder, while it is possible to perform the examples in this or any chapter in server mode, I recommend that you perform examples on your local computer to minimize any confusion with other individuals who may be utilizing JMP Clinical on the server. See Section 1.5 for a brief description of working within a server environment.

For snapshot comparisons, I have simulated a data snapshot for the **Nicardipine** trial by removing data from all data sets that occurs at or after 01AUG1989 [1]. This snapshot excludes several participants from the complete trial, and truncates the data for several other patients. Further, the ADSL data set is not available for this study to mimic the likely scenario where this data set may lag behind the SDTM data due to the derived population flags and efficacy endpoints that may be included as part of the data set. To register **Nicardipine Early Snapshot**, go to **Studies > Add Study from Folders > Settings > Load > NicardipineEarlySnapshot > OK** to load the sample setting (Figure 6.1 on page 201). Click **Run**. This will add a study called **Nicardipine Early Snapshot** to the **Current Study** drop down menu in the Clinical Starter, as well as the **Study** drop downs on all clinical dialogs.

Figure 6.1 Add Study Nicardipine Early Snapshot

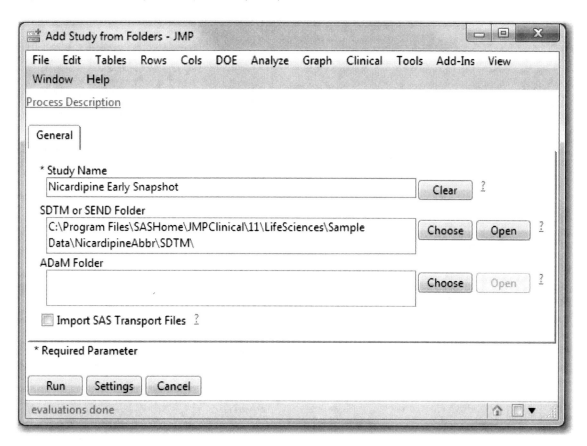

This, however, is only the first step for snapshot comparisons, since we have only just registered the initial snapshot. Once the study is added, a window will open with a button to **Open**

Duplicates and Keys Report; clicking this button will open a PDF file that describes the keys for all domains registered as part of the study (Figure 6.2 on page 202). This begs the question: what exactly are the keys? Keys typically give us access to something, whether it is a locked box or a room. When discussing data sets, keys give us insight into the uniqueness of a record or row within the data set. Section 3.2.1.1 of the CDISC SDTM Implementation Guide [2] defines the following terms:

1 *Natural Key* is one or more variables whose contents uniquely distinguish every record (row) in the data set. For example, each row of the DM domain should represent a different subject. The natural keys in this instance could be Study Identifier (DM.STUDYID) and Unique Subject Identifier (DM.USUBJID).

2 *Surrogate Key* is an artificially established single-variable identifier that uniquely identifies rows. This could include any of the xx.xxSEQ variables. For example, if the vital signs (VS) data set contained 200 records, the VS.VSSEQ variable could be numbered 1 to 200 to uniquely identify the rows.

Figure 6.2 Results for Add Nicardipine Early Snapshot

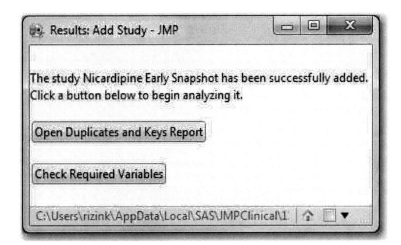

So why are keys important? In order to examine a data set record for differences between any two adjacent snapshots, there needs to be a way to link these two versions of the record together. Keys are this link. Otherwise, JMP Clinical has no way to know which records to match together from one snapshot to the next. Further, as I describe in Section 6.4, in order to save or access notes for a particular record, there needs to be a manner in which to file the note so that it is accessible later on. The keys for **Nicardipine Early Snapshot** (and by extension **Nicardipine**) from the **Duplicates and Keys Report** are listed in Table 6.1 on page 203.

Table 6.1 Keys for Nicardipine and Nicardipine Early Snapshot

Domain	Keys
AE	STUDYID USUBJID AETERM AESTDTC
CM	STUDYID USUBJID CMTRT CMSTDTC
DM	STUDYID USUBJID
DS	STUDYID USUBJID DSTERM DSSTDTC
EG	STUDYID USUBJID EGCAT EGTESTCD VISITNUM EGDTC
EX	STUDYID USUBJID EXSTDTC
LB	STUDYID USUBJID LBTESTCD VISITNUM LBDTC
MH	STUDYID USUBJID MHTERM MHSTDTC
SV	STUDYID USUBJID VISITNUM
VS	STUDYID USUBJID VSTESTCD VISITNUM VSDTC VSSEQ
ADSL*	STUDYID USUBJID

Note: * **Nicardipine** only

A user can easily define keys and provide them to JMP Clinical. Notice in the output below from PROC CONTENTS for the DM domain that the data set is sorted (Output 6.1 on page 204). Since the data is sorted (i.e., Sorted YES), Sort Information is provided in the output after the description of the data set variables (i.e. Sortedby DM.STUDYID DM.USUBJID). This metadata is stored in the SAS formatted data set, and the variables used for sorting the data set is what JMP Clinical uses to define the keys for a study.

Output 6.1 *Output for PROC CONTENTS of DM Domain*

```
The CONTENTS Procedure
Data Set Name         PERM.DM                      Observations           906
Member Type           DATA                         Variables              19
Engine                V9                           Indexes                0
Created               11/30/2012 15:00:05          Observation Length     288
Last Modified         11/30/2012 15:00:05          Deleted Observations   0
Protection                                         Compressed             NO
Data Set Type                                      Sorted                 YES

              Alphabetic List of Variables and Attributes

 #    Variable    Type    Len    Label

11    AGE         Num      8     Age
12    AGEU        Char    10     Age Units
16    ARM         Char    20     Description of Planned Arm
15    ARMCD       Char    20     Planned Arm Code
10    BRTHDTC     Char    20     Date/Time of Birth
17    COUNTRY     Char     3     Country
18    DMDTC       Char    20     Date/Time of Collection
19    DMDY        Num      8     Study Day of Collection
 2    DOMAIN      Char     2     Domain Abbreviation
 8    INVID       Char    10     Investigator Identifier
 9    INVNAM      Char    10     Investigator Name
14    RACE        Char    50     Race
 6    RFENDTC     Char    20     Subject Reference End Date/Time
 5    RFSTDTC     Char    20     Subject Reference Start Date/Time
13    SEX         Char     1     Sex
 7    SITEID      Char    10     Study Site Identifier
 1    STUDYID     Char     7     Study Identifier
 4    SUBJID      Char    40     Subject Identifier for the Study
 3    USUBJID     Char     6     Unique Subject Identifier

              Sort Information

Sortedby         STUDYID USUBJID
Validated        YES
Character Set    ANSI
```

To save sorting metadata to a SAS-formatted data set, the programming team can apply either of the following approaches:

```
PROC SORT data = DM out = library.DM;
   by STUDYID USUBJID;
run;
```

or

```
data library.DM(sortedby = STUDYID USUBJID);
   set DM;
```

```
run;
```

If study domains do not have the SORTEDBY metadata defined, JMP Clinical attempts to derive the keys based upon the presence of variables suggested in the SDTM Implementation Guide. Note that keys derived in this fashion may not represent the optimal set of variables for a given domain. This raises an important point: what happens if the supplied keys do not define the records uniquely within each domain? When the study is first added to JMP Clinical, the **Duplicates and Keys Report** details the records that cannot be uniquely determined based on the provided or derived keys. These records, like all records when first registering a study, will initially be labeled as New ("6.3.1 Record-Level" on page 212) in JMP Clinical. However, any record-level notes that are system- or user-generated for a particular set of key values where duplicates are present would naturally be associated with two or more records. This may be acceptable if there are few duplicates with which to contend, but any duplicates should be reviewed as potential data errors (data that was mistakenly entered twice) or deficiencies with the supplied keys. When the study data is updated and redundancies remain, JMP Clinical will be unable to match these records between the snapshots. These records are referred to as Non-Unique, since JMP Clinical cannot assess whether any changes occurred between the snapshots. Again, if there are few duplicates, these Non-Unique records can be reviewed at every snapshot for correctness. For **Nicardipine Early Snapshot**, you will notice that several duplications exist within the concomitant medications (CM) domain (Figure 6.3 on page 206). This is intentional for illustration.

Figure 6.3 Duplicates for CM Domain

CM: Keys with Duplicates (no comparisons were made with previous data if CMRFlg = Non-Unique)

Study Identifier	Unique Subject Identifier	Reported Name of Drug, Med, or Therapy	Start Date/Time of Medication	CM Review Flag	New Snapshot Frequency
NICSAH1	11008	INSULINS	1988-01-13T19:00:00	New	2
NICSAH1	121003	PHENOBARBITAL	1988-12-22T21:00:00	New	2
NICSAH1	122002	PHENOBARBITAL	1988-08-17T22:00:00	New	2
NICSAH1	141063	INSULINS	1989-01-01T00:30:00	New	2
NICSAH1	141071	MAGNESIUM HYDROXIDE	1989-03-17T18:30:00	New	2
NICSAH1	142003	GENTAMICIN	1989-04-28T14:00:00	New	2
NICSAH1	161001	DEXAMETHASONE	1988-04-21T06:00:00	New	2
NICSAH1	171006	BENZODIAZAPINE	1988-04-05T13:30:00	New	2
NICSAH1	171006	HYDRALAZINE	1988-04-09T16:00:00	New	2
NICSAH1	172001	AMINDOAPROIC ACID	1988-01-28T18:30:00	New	2
NICSAH1	172001	DEXAMETHASONE	1988-02-08T12:45:00	New	2
NICSAH1	172003	PHENOBARBITAL	1988-07-20T03:00:00	New	2
NICSAH1	181002	MORPHINE	1988-05-06T09:00:00	New	2
NICSAH1	181004	MEPERIDINE	1988-06-09T16:55:00	New	2
NICSAH1	181004	MEPERIDINE	1988-06-09T23:40:00	New	2
NICSAH1	181004	MEPERIDINE	1988-06-10T06:00:00	New	2
NICSAH1	181004	PHENOBARBITAL	1988-06-09T12:00:00	New	2
NICSAH1	181004	PHENOBARBITAL	1988-06-09T13:00:00	New	2
NICSAH1	181004	PHENOBARBITAL	1988-06-09T14:00:00	New	2
NICSAH1	181005	FUROSEMIDE	1988-06-14T15:00:00	New	2
NICSAH1	181007	CODEINE COMPOUND 1/2	1988-07-29T08:30:00	New	2
NICSAH1	181009	RANITIDINE	1988-08-13T06:00:00	New	2
NICSAH1	181009	RANITIDINE	1988-08-19T06:00:00	New	2
NICSAH1	181012	DEXAMETHASONE	1988-08-29T06:00:00	New	2
NICSAH1	181012	DOPAMINE	1988-08-26T14:00:00	New	2
NICSAH1	181012	PHENOBARBITAL	1988-08-29T06:00:00	New	2
NICSAH1	181014	DEXAMETHASONE	1988-09-03T06:00:00	New	2
NICSAH1	181014	MULTIVITAMINS	1988-09-09T10:00:00	New	2
NICSAH1	181015	DEXAMETHASONE	1988-10-23T06:00:00	New	2
NICSAH1	181015	MULTIVITAMINS	1988-10-23T10:00:00	New	2
NICSAH1	181015	MULTIVITAMINS	1988-10-26T10:00:00	New	2
NICSAH1	181015	PHENOBARBITAL	1988-10-23T12:00:00	New	2
NICSAH1	181015	PHENOBARBITAL	1988-10-26T10:00:00	New	2
NICSAH1	191001	PHENOBARBITAL	1988-05-23T06:00:00	New	2
NICSAH1	191006	LABETALOL HCL	1989-06-17T10:00:00	New	2

Below are some important points to consider in order to effectively use the snapshot comparison feature of JMP Clinical.

1. When the study is first registered within JMP Clinical using one of the **Add Study** reports, be sure to examine the **Duplicates and Keys Report**. Identify the keys for each domain and make sure any duplicates are kept to a minimum. Ideally, duplicates will not be present. Otherwise, the snapshot comparisons functionality will not be as useful as it ultimately could be. For example, if the vital signs (VS) domain was sorted only by xx.STUDYID and xx.USUBJID using the PROC SORT code above, all records for each subject would be considered duplications. These duplicates would include the multiple tests present (such as heart rate, systolic, and diastolic blood pressures) as well as records belonging to different visits. This is hardly useful for our purposes.

2. Plan your keys considering the whole study. For example, if there are only two records in the VS domain, one each for two different subjects, xx.USUBJID alone would be sufficient to distinguish these two records. However, once additional data are collected, xx.USUBJID will likely become insufficient to serve as a key alone. Let the study design dictate appropriate choices for keys, and use the SDTM Implementation Guide for suggestions.

3. If there are numerous duplications after registering the study as described above, use **Studies > Manage Studies > Remove Study** to unregister the study from JMP Clinical. Re-register the study once more appropriate keys have been defined for the data sets by the programming team. If it takes a long time to register the study in JMP Clinical, this may be an indication that numerous duplications among records exist.

4. Choose the smallest number of variables possible to define the keys, and choose variables that are not likely to change values. If changes do occur for one or more variables that comprise the keys, there is currently no way to match these record(s) to previous versions of the record(s) to observe changes over time or any notes that were defined for these records. However, since all data set records have Unique Subject Identifier (xx.USUBJID) available as part of the keys, it is possible to view all notes at the subject level. More details for notes can be found in Section 6.4. Use the SDTM Implementation Guide for recommendations for keys.

5. Given item 4, I would suggest not using terms that rely on medical coding as keys (e.g., AE.AEDECOD based on MedDRA or CM.CMDECOD based on WHODRUG). There are two reasons for this. First, medical coding may not be immediately available. This provides an opportunity for a missing value of AE.AEDECOD to change to a non-missing coded term later on. Second, sometimes over the course of a study, coded terms may change based on new insights of the clinical team. I would recommend using verbatim terms such as AE.AETERM or CM.CMTRT. However, if the team is diligent in coding terms as they are added to the study, using coded variables xx.xxDECOD is probably okay in practice.

6. The xx.xxSEQ variable or the xx.STUDYID, xx.USUBJID and xx.xxSEQ set may be good keys to use since these values are unlikely to change. HOWEVER, the xx.xxSEQ variable must be carefully maintained so that the number never changes for a particular record. For example, suppose a CM data set contains two records (Table 6.2 on page 208), and is updated through query with a new med that actually falls between the first two based on date (Table 6.3 on page 208). It is important that any new records are tacked at the end (and to continue the sequence of CM.CMSEQ). Alternatively, if a record is deleted as in Table 6.4 on page 208,

the sequence number must be kept consistent (i.e., 1 can never be used again). If your company tends to define xx.xxSEQ as 1 to *N* for all records or 1 to n_i for each subject without any concern for what the row values are, using xx.xxSEQ as a key is not a good choice.

7. Alternatively, a single non-CDISC variable can be included in each domain and added to the SORTEDBY metadata. A good choice may include a record-identifier variable output from any DBMS, but this would likely require the DBMS to be formatted to handle CDISC data structures. These non-CDISC keys can be removed later on in order to submit the trial data to regulatory agencies. Using a single non-CDISC variable as a key can help avoid issues regarding Non-Unique matches between snapshots.

Table 6.2 Concomitant Medication Example

CM.CMSEQ	CM.CMTRT	CM.CMSTDTC
1	ASPIRIN	03-20-1974
2	IBUPROFEN	03-27-1974

Table 6.3 Concomitant Medication Example with New Record

CM.CMSEQ	CM.CMTRT	CM.CMSTDTC
1	ASPIRIN	03-20-1974
2	IBUPROFEN	03-27-1974
3	VITAMIN C	03-24-1974

Table 6.4 Concomitant Medication Example with Record Deleted

CM.CMSEQ	CM.CMTRT	CM.CMSTDTC
2	IBUPROFEN	03-27-1974
3	VITAMIN C	03-24-1974

TIP When registering a study within JMP Clinical, review the **Duplicates and Keys Report** to assess whether domain keys are defined appropriately. Incorrectly-defined keys can limit the effectiveness of the snapshot comparison feature.

If there are no duplicates present using the selected keys or if there are an acceptable number of duplications (such as for **Nicardipine Early Snapshot**) in the **Duplicates and Keys Report**, proceed to analyze and review the data accordingly. When a new snapshot becomes available, go to **Studies > Manage Studies > Update Study Data and Metadata**, provide the new **SDTM** and/or **ADaM Folders** for the study data sets, and click **Run**. For **Nicardipine Early Snapshot**, go to **Studies > Manage Studies > Settings > Load > NicardipineEarlySnapshot > OK** to update to the next snapshot (Figure 6.4 on page 210). Click **Run**. This "new" snapshot happens to be the full set of SDTM data sets for the **Nicardipine** study. If selected, the option **Exclude comparisons of treatment variables** will not perform any comparisons between treatment variables in DM, ADSL or EX. This option exists to minimize any unnecessary output among treatment variables that may occur from using dummy or blinded treatment codes and updating to actual treatment codes.

Figure 6.4 *Manage Studies for Nicardipine Early Snapshot*

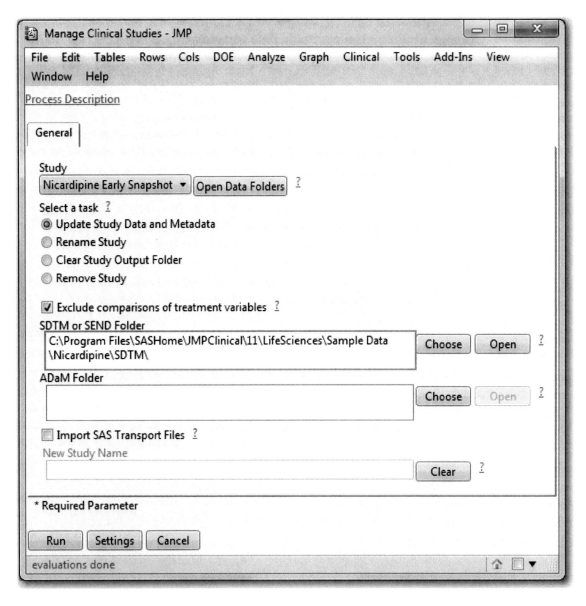

Once JMP Clinical has updated metadata, a window will open stating that the current review number is 1 (Figure 6.5 on page 211; the review number is 0 when studies are first added). A button to **Open Differences Report** will open a PDF document that summarizes Non-Unique records where no comparisons were performed, records where changes occurred from the previous snapshot to the current one, as well as records that were Dropped. For **Nicardipine Early Snapshot**, there are changes in records from the DM, AE, CM, and EX domains, mostly due to the availability of an end date. Further, the CM domain has Non-Unique records where no comparisons were performed. If an additional snapshot became available for **Nicardipine Early**

Snapshot, we would repeat the process to **Update Study Data and Metadata** as was done in the previous paragraph.

Figure 6.5 Results of Manage Studies for Nicardipine Early Snapshot

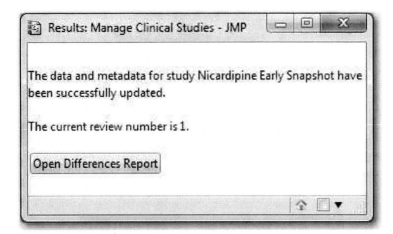

It is important to note that JMP Clinical expects the supplied directories to be different from one snapshot to the next, and in case any problems occur in the data, it is generally good practice to maintain these directories to identify the point at which problems arose. Replacing old data sets with updated data sets and using the same directory for new snapshots will cause JMP Clinical to perform comparisons of the new snapshot against itself.

Finally, snapshot comparisons are performed between adjacent snapshot pairs as presented in Figure 6.6 on page 212. In general, comparisons between any two arbitrary snapshots are not performed in JMP Clinical, though enterprising users can register a new study and apply snapshots in the appropriate order to obtain the desired direct comparisons. As an alternative to this approach, the review of system-generated notes (described below) will illustrate changes in a record over multiple snapshots to describe its full history.

Figure 6.6 How Snapshot Comparisons are Performed

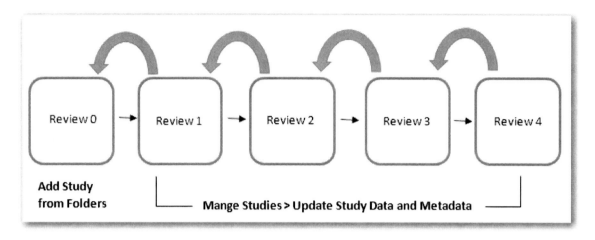

6.3 Review Flags

6.3.1 Record-Level

The comparisons between snapshots are summarized in two ways. The first comparison is at the record level, providing details for every distinct set of key values within each domain. The second set of flags summarizes the record-level flags at the patient level; these flags will be described in the next section. These review flags only describe the relationship between the previous and current data snapshots; they are not intended to describe the history of a record over the course of the clinical trial. System-generated notes can help describe how a record changes over time. Note that the record-level review flags are saved independently of the study database. In other words, no modifications are made to the original domain data sets. The available record-level review flags are summarized in Table 6.5 on page 213, and are based on the particular set of key values between the current and previous snapshot. Note that while ADSL was not included in the second data snapshot for **Nicardipine Early Snapshot**, it easily could have been. In this case, since it wouldn't have been available in the previous snapshot, all record-level flags would be 'New.'

Table 6.5   Record-Level Review Flags

Record-Level Review Flag	Details
New	Indicates a record that was not available in the previous snapshot. When a study is first registered with JMP Clinical, all records are considered New, even those records considered duplicates based on the current keys.
Stable	Indicates a record that had no changes between the previous and current snapshot.
Modified	Indicates a record that had changes in at least one variable between the previous and current snapshot.
Non-Unique	Indicates a record that was not compared between the previous and current snapshots due to duplications among key values. This duplication may exist in either the previous snapshot only, the current snapshot only, or both. Records labeled as Non-Unique in the current snapshot but present as a single record imply the keys are duplicated in the previous snapshot. Records labeled as Non-Unique in the current snapshot and present as multiple records could imply that keys are duplicated in the previous snapshot or that the duplication is new. Record-level notes can help distinguish between these cases.
Dropped	Indicates a record that was available in the previous snapshot but not the current snapshot. This may be due to changes in key values for a particular record. In general, this review flag is only visible using a drill down in **Studies > Domain Viewer** since these records are not available in the current snapshot.

Record-level review flags are SAS variables of the form xxRFlg and the label "XX Review Flag". For ADSL, the ADSLRFlg variable has the label "ADSL Review Flag".

6.3.2 Patient-Level

The patient-level review flags summarize the record-level flags at the patient level, though these flags do not account for the presence of any changes that may or may not be occurring between Non-Unique records, or records that have Dropped since the previous database. In this case, users should **Open Differences Report** after using **Manage Studies** to update data to determine if there are any subjects where review flags may not be wholly accurate. Domains with numerous duplications can be reviewed using **Studies > Domain Viewer**. Note that these patient-level review flags only describe the relationship between the previous and current data snapshots, and do not describe the history of a record over the course of the clinical trial. System-generated notes can help describe how records for a study participant change over time. Note that the patient-level review flags are saved independently of the study database. In other words, no modifications are made to the original domain data sets. The available patient-level review flags are summarized in Table 6.6 on page 214, and are based on the current set of record-level flags for each Unique Subject Identifier (xx.USUBJID). The patient-level review flag is a SAS variables called ReviewFlag with the label "Review Flag".

Table 6.6 Patient-Level Review Flags

Patient-Level Review Flag	Details
Stable	Indicates a participant with no New or Modified records.
New Records	Indicates a participant that has at least one New record of data available. This individual may also have Modified records.
Modified Records	Indicates a participant that has no New records, but has had changes in at least one record between the previous and current snapshot.
Introduced	Indicates a participant that was not present in the previous snapshot. All of their data is considered New.

The patient-review flags can be used to create data filters to subset to those individuals with New or Modified data. Now that **Nicardipine Early Snapshot** has been updated to the most complete set of SDTM data, let's examine the data using **Subject Utilities > Review Status Distribution**. In earlier releases of JMP Clinical, the **Review Status Distribution** report was a way for reviewers to easily identify patient profiles that had yet to be reviewed. This report also serves as a place to identify and define patient subsets based upon their review flags. Click **Run** to generate output for **All Subjects Excluding Screen Failures** (see the **Filter** tab) for **Nicardipine Early Snapshot**. In Figure 6.7 on page 215, the patient profile Review Status, the patient-level

Review Flag, and the presence of system- or user-defined notes are summarized in the histograms. Of most interest to us is the histogram for Review Status. We can immediately observe that 861 study participants (95.5% of those not considered screen failures) have no changes to their data between the current and previous snapshot. This is a huge advantage to reviewers who wish to exclude patients with no new data to examine. Click on **AND** in the **Data Filter** to add Review Flag, since it will make it much easier to select subjects with New or Modified data (Figure 6.8 on page 216). Subset the histograms to those subjects not considered Stable.

Figure 6.7 Review Status Distribution

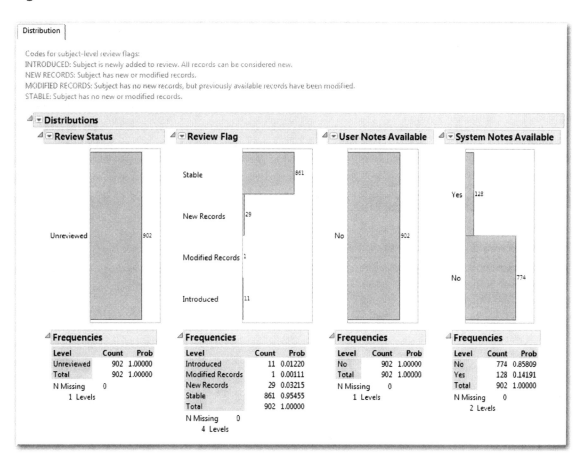

Figure 6.8   Data Filter with Patient-Level Review Flag

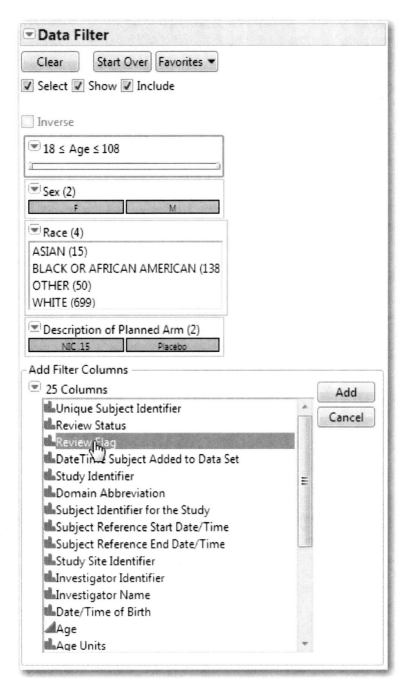

Of these 41 individuals, 11 are newly-available in this snapshot, 29 have additional records to review, and a single individual has at least one modification to records that existed in the previous snapshot (e.g., perhaps an ongoing adverse event finally resolved). Click the **Create Subject**

Filter drill down, name this filter **New Data,** and click **OK** (Figure 6.9 on page 217). This filter will allow the analyst to subset most analyses to the data of these 41 individuals using the **Subject Filter** dropdown box on the **Filters** Tab of report dialogs. Note that applying these filters does nothing to subset the records for these individuals to those that may be New or Modified, but we have saved enormous effort by removing subjects with no New or Modified data. I show how to incorporate record-level review flags into analyses and reports in "6.5 Using Review Flags" on page 220.

Figure 6.9 Naming a Subject Filter

6.4 Adding and Viewing Notes

The **Add Notes** and **View Notes** drill downs are available in most JMP Clinical analyses and reports (Figure 6.10 on page 218). These features permit the user to create and view notes on any unusual findings or trends that may be important to recall later on, such as to include these details in the clinical study report. Prior to JMP Clinical 5.0, notes could be made at the analysis, subject, or record level. Beginning with JMP Clinical 5.0, the user can additionally create notes for some reports in the **Risk-Based Monitoring** menu at the site and country level. The type of notes that are created depends on the particular report, and whether or not data are selected in the currently active data table. Data are typically selected as a result of selecting various features in the report or subsetting results using a **Data Filter**. If using JMP Clinical in local mode, all notes are limited to those of a single individual on their computer. If working in server mode, any notes made using **Add Notes** will be available to anyone with access to the particular study. System notes are also generated and available to inform the user whether records associated with a particular domain and key values have been dropped or duplicated, or whether a modification is present in one or more variables. Notes are date-time stamped at the time they are generated. The types of notes available are described in Table 6.7 on page 218.

Figure 6.10 Add Notes and View Notes Drill Downs

Table 6.7 Notes Available Within JMP Clinical

Note Type	Details
Analysis	Notes can be created at the analysis or report level. To do this, no data can be selected from any data table. When **Add Notes** is clicked, the notes window will say **Enter a Note for the Current Process Output**. Notes are filed using the name of the report. To view analysis-level notes, click **View Notes** while no data are selected in the currently active data table. This will also show patient- or record-level notes created within the report. Analysis-level notes are missing values for Unique Subject Identifier (xx.USUBJID), Keys, and Keys Values.
Patient	Notes can be created at the patient or subject level. When **Add Notes** is clicked while rows from a data table that includes Unique Subject Identifier (xx.USUBJID) are selected, the notes window will say **Enter a Note for Selected Subjects**. Notes are filed using xx.USUBJID and the name of the report. To view notes for particular subjects, click **View Notes** while these individuals are selected. This will also show all record-level notes available for these subjects. Notes made in the patient profiler are made at the patient level. Patient-level notes are missing values for Keys and Keys Values.

Note Type	Details
Record	Notes can be created at the record level using **Studies > Domain Viewer** (Section 6.5.1). When **Add Record-Level Notes** is clicked while rows from a data table are selected, the notes window will say **Enter a Note for Selected Records**. Notes are filed using xx.USUBJID, the Keys and their values, the CDISC domain, and the name of the report. Record-level notes are also generated by the system to describe records that have been dropped or duplicated, or whether a modification is present in one or more variables for a particular record. System-generated record-level notes has JMP Clinical listed as the User. To view notes for particular records, click **View Record-Level Notes** while these records are selected. Clicking **View Domain Notes** shows all record-level notes for the selected domain. Record-level notes have non-missing values for Keys and Keys Values.
Country	Notes can be created at the country level for certain reports within the **Risk-Based Monitoring** menu (Chapters 2-3). When **Add Notes** is pushed while rows from a data table are selected that includes Country (DM.COUNTRY) but not Unique Subject Identifier (xx.USUBJID) or Study Site Identifier (DM.SITEID), the notes window will say **Enter a Note for Selected Countries**. Notes are filed using DM.COUNTRY and the name of the report. To view notes for particular countries, click **View Notes** while these countries are selected. Country-level notes have non-missing values for Country.

Note Type	Details
Site	Notes can be created at the site level for certain reports within the **Risk-Based Monitoring** menu (Chapters 2-3). When **Add Notes** is pushed while rows from a data table are selected that includes Study Site Identifier (DM.SITEID) but not Unique Subject Identifier (xx.USUBJID), the notes window will say **Enter a Note for Selected Sites**. Notes are filed using DM.SITEID and the name of the report. To view notes for particular sites, click **View Notes** while these countries are selected. Site-level notes have non-missing values for Study Site Identifier (DM.SITEID).

6.5 Using Review Flags

6.5.1 The Domain Viewer

In this section, I describe several ways in which the user can take advantage of the review flags that the snapshot comparisons feature makes available for analysis. For example, the **Domain Viewer**, available from the **Studies** menu, provides a more traditional way to review the clinical trial in that the individual domains are presented as JMP data tables. Similar to a spreadsheet, data can be sorted and additional variables created in order to review the data for any individual subject. Unlike typical spreadsheets, however, the power and flexibility of JMP is available at your fingertips to further **Analyze** and **Graph** the data. There are numerous other benefits as well. From the **Domain Viewer** dialog, it is possible to subset data using the **Select Analysis Population** or **Subject Filter** drop-down menus. These options make it straightforward to initially subset all domains to subjects belonging to the Safety Population, subsets of patients meeting specific demographic criteria, or to those individuals with new or modified data by applying the New Data filter that was created in Section 6.3.2. For now, **Run** the dialog for **Nicardipine Early Snapshot** for All Subjects Excluding Screen Failures.

Figure 6.11 The Initial View of the Domain Viewer

Initially, the dashboard for **Domain Viewer** does not display any data (Figure 6.11 on page 221). **Domains** can be selected from the Domains drop-down menu and opened by clicking the **View Domain** drill down. Open the DM data table; the 902 rows indicate that there are 902 patients not currently considered screen failures for the **Nicardipine Early Snapshot** study (Figure 6.12 on page 222). The colored row markers indicate an additional benefit of the **Domain Viewer** beyond standard spreadsheets in that they are a visual cue for the record-level review flags described in "6.3.1 Record-Level" on page 212. Table 6.8 on page 222 describes the color for each record-level review flag.

Figure 6.12   Domain Viewer with DM Data

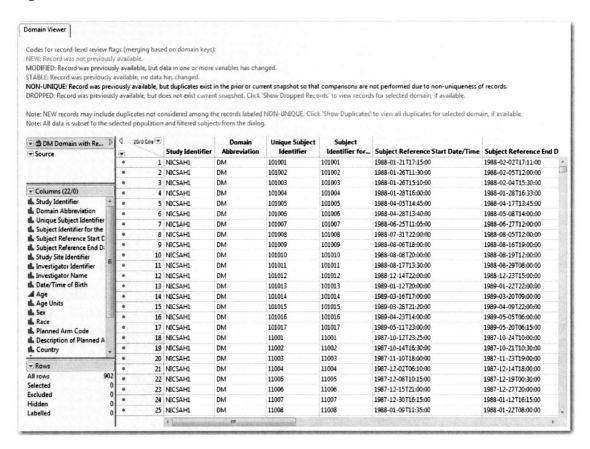

Table 6.8   Row Marker Colors for Record-Level Review Flag Status

Record-Level Review Flag	Color
New	Yellow
Stable	Green
Modified	Red
Non-Unique	Black
Dropped	N/A, since the dropped records are not available in the domains

Figure 6.13 *Examples of Varying Review Flags and Row Marker Colors*

	Study Identifier	Domain Abbreviation	Unique Subject Identifier	Subject Identifier for...	Subject Reference Start Date/Time	Subject Reference End Date/Time	Study Site Identifier
●	580 NICSAH1	DM	321037	321037	1989-06-19T15:00:00	1989-06-30T06:00:00	32
●	581 NICSAH1	DM	321038	321038	1989-06-30T07:45:00	1989-07-11T08:00:00	32
●	582 NICSAH1	DM	321039	321039	1989-07-06T07:45:00	1989-07-14T17:30:00	32
✳	583 NICSAH1	DM	321040	321040	1989-07-22T12:25:00	1989-08-02T09:00:00	32
✳	584 NICSAH1	DM	321041	321041	1989-07-29T11:15:00	1989-08-04T12:30:00	32
✳	585 NICSAH1	DM	321042	321042	1989-07-31T10:45:00	1989-08-12T16:00:00	32
●	586 NICSAH1	DM	321043	321043	1989-08-07T15:30:00	1989-08-25T10:00:00	32
●	587 NICSAH1	DM	321044	321044	1989-08-07T12:15:00	1989-08-16T10:15:00	32
●	588 NICSAH1	DM	322001	322001	1989-04-21T21:00:00	1989-05-03T19:00:00	32

These filled-circle markers make it easy to identify rows with New or Modified data, or those records that are Non-Unique to streamline review. Scroll down to row 580 to view examples of the other review flags (Figure 6.13 on page 223). Notice the records with red markers, which indicate rows that have one or more variables with modifications since the last snapshot. The variables with modifications, here only Subject Reference End Date/Time (DM.RFENDTC), have their cells highlighted in red. While the red color indicates variables that have changes associated with them, it does not describe the actual modification. The filled-circle marker has been replaced by an asterisk to indicate a row that has record-level notes associated with it. Record-level notes are often system-generated to describe a duplication, or in this case, modifications to the variables within the record. Select the three rows with red markers and click **View Record-Level Notes**. A data table of notes (one per subject per modified variable) tells us that the reference end dates were missing in the previous snapshot, most likely because the patients were still ongoing in the trial (Figure 6.14 on page 223). In this way, the analyst can view changes to a record across multiple snapshots. Note that red markers are always asterisks, since there will always be at least one note present to describe a modification. To view all record-level notes for this domain, click **View Domain Notes**.

Figure 6.14 *Record-Level Notes for Patients with a Data Modification*

	Unique Subject Identifier	Study Name	Keys	Keys Values of notes	Note	Date of Note
1	321040	Nicardipine Early Snapshot	STUDYID USUBJID	NICSAH1~321040	Value change in Subject Reference End Date/Time (RFENDTC) from Missing to 1989-08-02T09:00:00.	12Apr2014 5:58:30 P
2	321041	Nicardipine Early Snapshot	STUDYID USUBJID	NICSAH1~321041	Value change in Subject Reference End Date/Time (RFENDTC) from Missing to 1989-08-04T12:30:00.	12Apr2014 5:58:30 P
3	321042	Nicardipine Early Snapshot	STUDYID USUBJID	NICSAH1~321042	Value change in Subject Reference End Date/Time (RFENDTC) from Missing to 1989-08-12T16:00:00.	12Apr2014 5:58:30 P

While most record level notes are system-generated, the user is free to create their own record-level notes to describe unusual features in the data. For example, go to row 289 to identify the patient that is 108 years old. Select the row and click **Add Record-Level Notes** to add a note to indicate that this subject is an outlier in terms of their age (Figure 6.15 on page 224). This changes the green filled-circle marker to an asterisk to indicate that there is now a note available for this record (Figure 6.16 on page 224). Note that if multiple rows are selected when creating a note, an identical but separate note is created for each record. In the **Domain Viewer**, the **Add Notes** and **View Notes** drill downs will create and view subject-level notes, respectively, when rows of a data table are selected. When no rows are selected, notes are created or viewed at the analysis level. Viewing analysis-level notes includes any record-level or subject-level notes created by the user. Changing **Domains** will close the currently open data table.

Figure 6.15   *Record-Level Note for Row 289*

Figure 6.16   *Row Marker Indicates a New Note is Available*

	Study Identifier	Domain Abbreviation	Unique Subject Identifier	Age
289	NICSAH1	DM	21025	108

Other drill downs for **Domain Viewer** include:

1. **Show Duplicates** opens a data table that summarizes duplicate rows. Change the **Domain** to CM and use this drill down to view the duplicate records present for this domain. This includes records that are

    a. Non-Unique, which are duplicates that exist among keys in both the current and previous snapshot, and

    b. New, where there are multiple instances of keys are currently only present in the current snapshot.

2. **Show Domain Keys** opens a window that describes how the uniqueness of records is currently determined for the data table.

3. **Show Dropped Records** opens a data table of records from the previous snapshot that are not available in the current snapshot. While this may represent records that were truly deleted, it could indicate rows that have had changes among one or more variables that are keys (if this is the case, these records would be New). If needed, **View Notes** associated with Dropped records and add them to the appropriate New record. Alternatively, create a record-level note for the New record to be sure to view all notes associated with this patient (it is extremely unlikely that the Unique Subject Identifier would change). **Nicardipine Early Snapshot** has no Dropped records to view.

Finally, when the data tables are first opened, they are locked to prevent any inadvertent changes to the data within the table. To unlock the data table, open the left Tab next to the data table to display the menus, click the red triangle, and select **Lock Data Table**. The table can now have variables added, formulas applied, or sorts performed (Figure 6.17 on page 225).

Figure 6.17 Unlocking a Data Table in Order to Make Changes

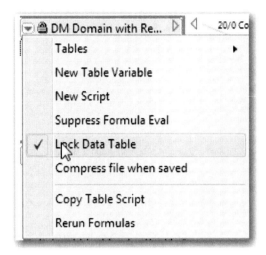

6.5.2 Demographic Distribution

Open **Demographics and Visits > Demographic Distribution**. On the **Filters** tab, select the New Data **Subject Filter** (Figure 6.18 on page 226). Click **Run** for **Nicardipine Early Snapshot**. The output (Figure 6.19 on page 227) summarizes the demography for the 41 individuals that were either introduced in this snapshot, have a modification in at least one record but no new data (Modified Records), or new records and or modifications (New Records). Note that while these subjects have new records or modifications, this may not necessarily have anything to do with the records summarized within **Demographic Distribution** (except for the subjects that were introduced, where everything is New). Now that the data for **Nicardipine Early Snapshot** has been updated at least once, the **Data Filter** includes the record-level review flag for DM (Figure 6.20 on page 228). This makes it possible to select records within the DM domain that are New, Modified, or Stable. Since demography is collected when first entering any clinical trial, selecting New in the **Data Filter** subsets to those 11 patients that are to the study new since the last snapshot. Here, 17 patients have no changes in demography (Stable) while 13 participants have one or more modifications to their demographic data. Clicking **View Notes** and sorting the data table by Domain show that all changes to DM involve Subject Reference End Date/Time (DM.RFENDTC), which is currently not summarized in the **Distribution Details** tab (though available with **View Data**). So, in actuality, only the 11 New records represent changes to demography presentation since the last snapshot. Subset to those records considered Modified and click **Patient Profiles**.

Figure 6.18 *Selecting Patients with New or Modified Data Using the Subject Filter*

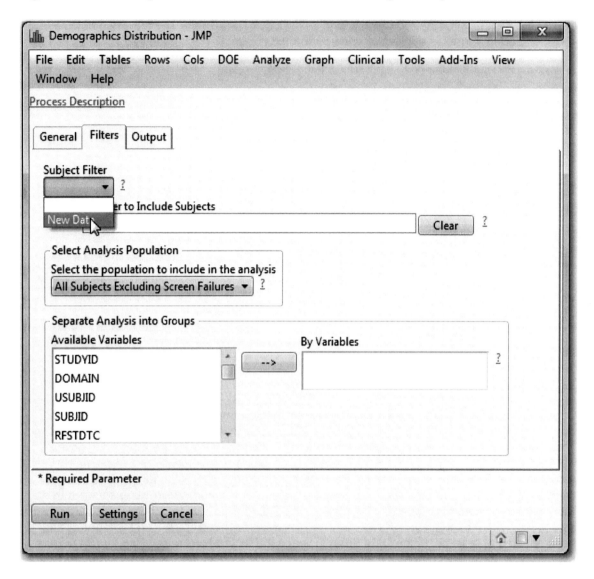

Figure 6.19 Demographic Results for Subjects with New or Modified Data

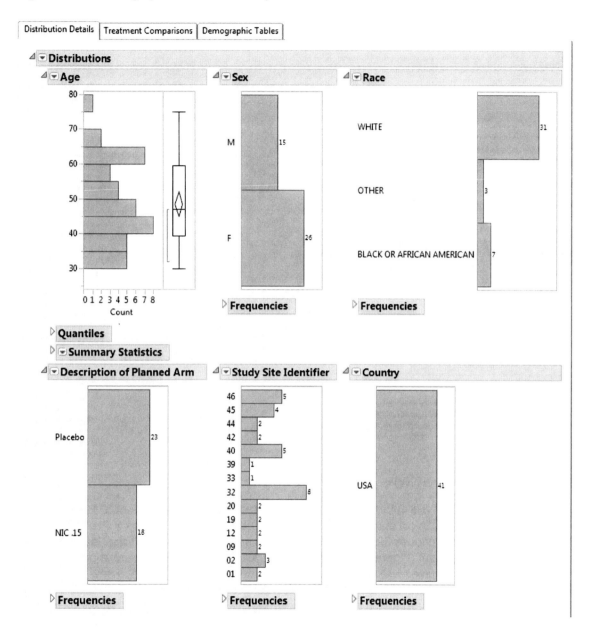

Figure 6.20 DM Record-Level Review Flag in the Data Filter

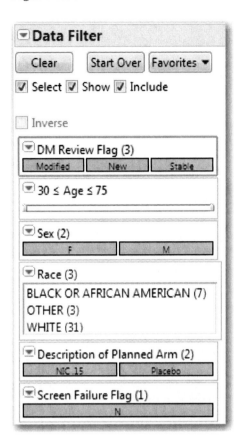

6.5.3 Patient Profiles

After performing the analysis in the previous section, select the profile for patient 321041. Alternatively, if no JMP Clinical output is currently open, go to **Subject Utilities > Profile Selected Subjects** and choose 321041 from the list, an individual whose participation spans both sides of the study snapshot (Figure 6.21 on page 229). We have reviewed example patient profiles before, but for studies where the data has been updated at least once, a new button becomes available in the left hand panel: **Show New or Modified Records** (Figure 6.22 on page 230). Selecting this button redraws the profile subsetting data to those records that are New, had a modification to at least one variable (Modified), or were unable to be compared due to duplications of the record in either the previous or current data snapshot (Non-Unique). This also subsets records in the tabular profile presented on the **Tables** tab; Stable records are excluded. The new graphical profile is presented in Figure 6.23 on page 231.

Figure 6.21   Profile Selected Subjects Dialog

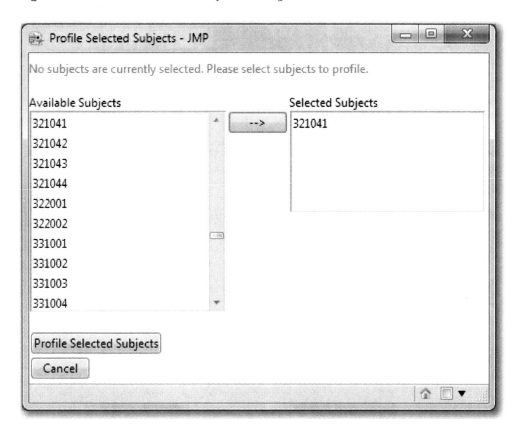

Data from Visits 1, 2, and 3 from records that remained unchanged have been removed from the profiler (based on absence of the symbols). This includes medical history terms, visit attendance, lab and vital sign results, and concomitant medications such as acetaminophen. Data from records that are New or Modified remain. To identify the specific modifications, the **View Notes** button shows all notes that were generated by JMP Clinical (Figure 6.24 on page 232). Most changes were due to adverse events or study medication use that was ongoing at the time of first snapshot, both of which have end dates available in the new snapshot. This explains why adverse events that began at earlier visits are still listed in the profile after choosing **Show New or Modified Records**. Again, this is an important point first described in the previous section; review-flags are at the record level. Changes for individual variable values for Modified records are available through notes. Non-Unique records due to duplicates (not an issue for this subject) would also be identified in this data table. All other values presented in the profile can be considered New for this subject. To return profiles to their previous state where all data was included, click **Show All Records**.

230 Chapter 6 / Snapshot Comparisons

Figure 6.22 Patient Profile with All Records Included

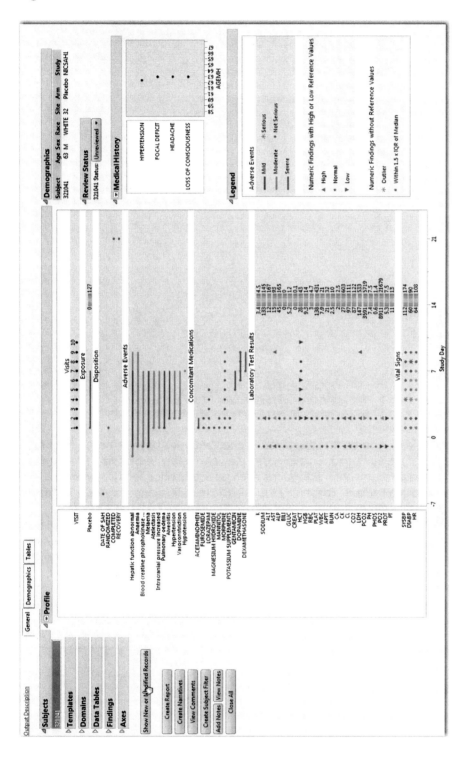

Figure 6.23 Patient Profile with New or Modified Records

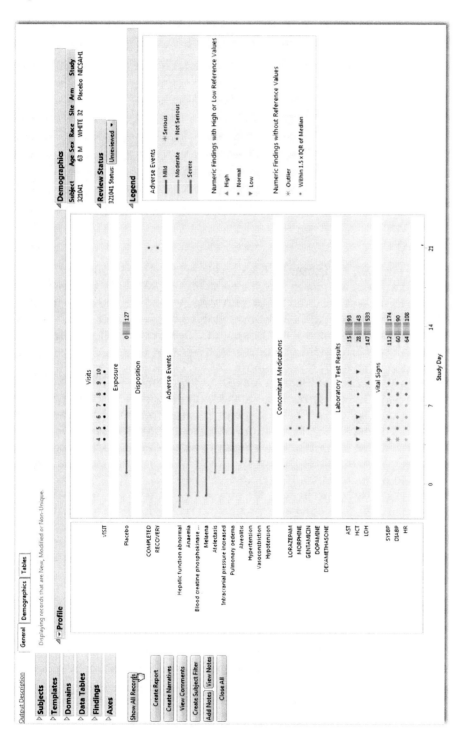

Figure 6.24   Notes Detailing the Modified Records

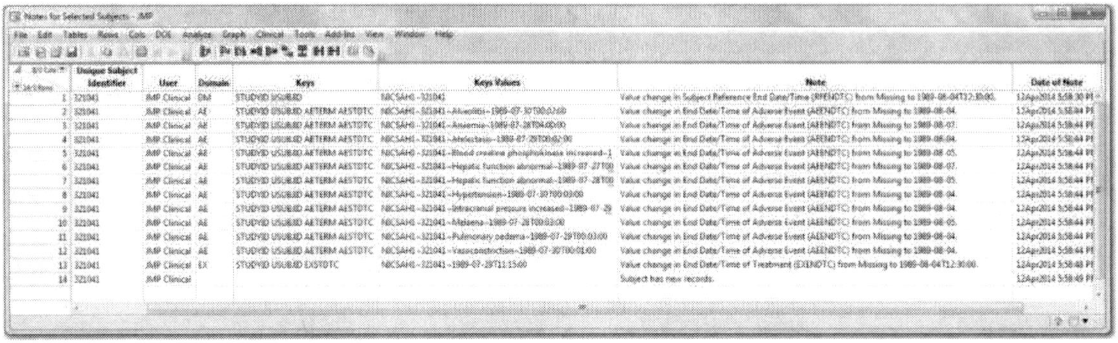

6.5.4 AE Distribution

Go to **Events > Adverse Events > AE Distribution**. Click on the option **Count multiple occurrences of an event per subject** in order to summarize all of the rows of the AE domain, otherwise the analyses collapse multiple occurrences of each term by choosing the earliest, most severe event (click the ? in the dialog for details). Make sure that the **Subject Filter** on the **Filters** tab is removed. **Run** the report for **Nicardipine Early Snapshot**. The report initially displays histograms of adverse events by the **Treatment or Comparison Variable to Use**. While Actual is selected in the dialog, the **Nicardipine** study does not have DM.ACTARM available; the treatment therefore defaults to Description of Planned Arm (DM.ARM). Similar to "6.5.2 Demographic Distribution" on page 225 for demographics, the **Data Filter** here includes a record-level review flag for AE so that the user can subset to displays to those events that are New, Modified, or Stable (Figure 6.25 on page 233). Further, click **AND** to add the DM Review Flag to the **Data Filter**. This provides the user with additional flexibility to select events, such as those AEs for subjects that are new to the trial (where DM Review Flag = "New"). **Clear** the **Data Filter** and subset to events where the Body System or Organ Class is CARDIAC DISORDERS. Click on AE Review Flag in the AE Stacking drill down to color the bars in the figure by the record-level review flag (Figure 6.26 on page 234). This can help describe differences in the distribution of events from the previous to the current snapshot to highlight any unusual patterns in safety. For this limited example, it allows us to see that no new event types occur for CARDIAC DISORDERS. Changing **Demographic Grouping** to None will group all events regardless of treatment.

Figure 6.25 AE Distribution with Review Flags

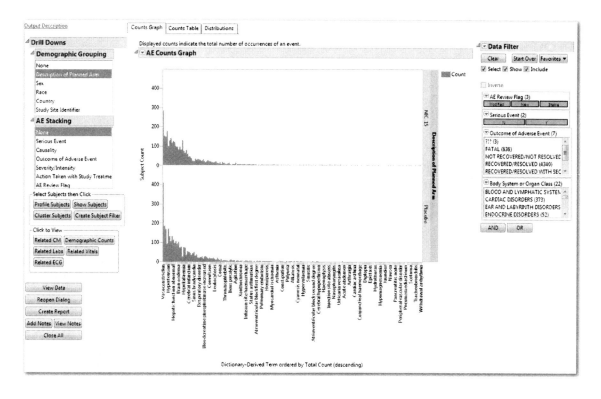

234 Chapter 6 / Snapshot Comparisons

Figure 6.26 Cardiac Disorders Colored by AE Record-Level Review Flag

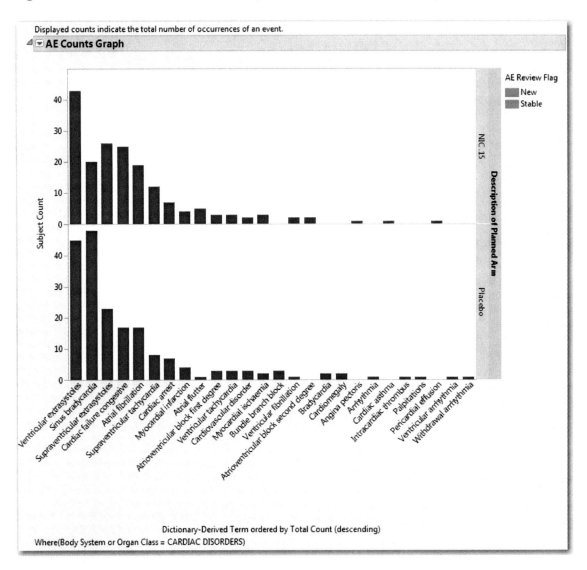

To illustrate a new addition to JMP Clinical 5.0, go to **Cols > Convert Character Date** (Figure 6.27 on page 235). Most date-time fields in CDISC are presented as character variables since this allows any partial information to be provided. While there are benefits to this approach, adding these dates to the JMP Data Filter means that we have to select specific dates. In other words, choosing a date range is less straightforward. Convert the Start Date/Time of the Adverse Event to create a numeric version of the date on which the AE began. This action adds a numeric version of the variable to the underlying data table. Click **AND** to add Numeric Start Date/Time of the Adverse Event to the **Data Filter**, which should be at the end of the list of variables (Figure 6.28 on page 236). Events can now be subset to a specific date range of interest.

Figure 6.27 Convert Character Date

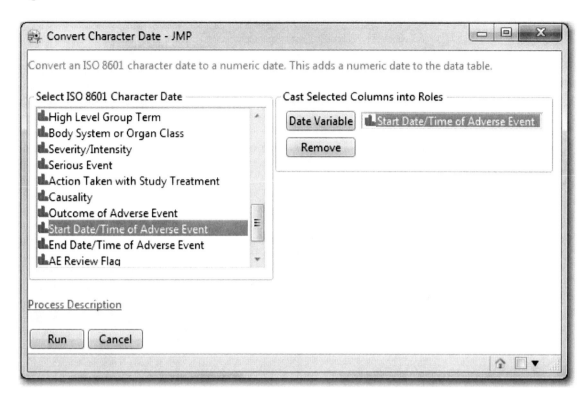

Figure 6.28 Numeric Date Added to Data Filter

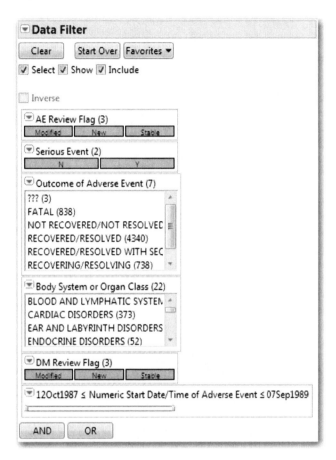

The last point about record-level review flags may initially be a bit confusing. Review flags are certainly useful in the data filter when summarizing individual records in a domain, such as **DM Distribution** or **AE Distribution** with **Count multiple occurrences of an event per subject** selected. When an analysis collapses over records in a domain, such as **AE Distribution** without the **Count multiple occurrences of an event per subject option** selected, the review-flags need to be used with more care. Why is this? In this instance, JMP Clinical chooses a single representative AE (in this case the earliest and most severe) to represent each AE type for each subject in the analysis, and this selection does not consider the AE Review Flag (New, Modified, Stable). When I filter the analysis in this case to New events, I am only subsetting to the New events that just so happened to be selected by the analysis. Other New events may be available that were not selected by the analysis since an earlier, more severe Stable event may have been selected.

How is it possible to run the default AE analysis, but only consider New events in this analysis when selecting the earlier, most severe AEs? First, deselect **Count multiple occurrences of an event per subject**. On the **Filters** Tab for **Include the following adverse events**, select New (Figure 6.29 on page 237). This will subset AEs to New events while also applying the **Filter to Include Adverse Events**, and **Event Type** on the **General** tab. The analysis will then choose the earliest, most severe (New) event for the report. To consider all events, change **Include the following adverse events** back to All. A similar effect to subset to New events could be obtained by writing aerflg = "New" in the **Filter to Include Adverse Events**. As another example presented in Figure 6.30 on page 237, type aerflg ne "New" (where *ne* means "not equal") or aerflg in ("Stable","Modified") to limit the analysis to AEs that were previously available.

Figure 6.29 Using the Record-Level Flag to Subset to New Events

Figure 6.30 Using the Where Statement to Subset to Previous Events

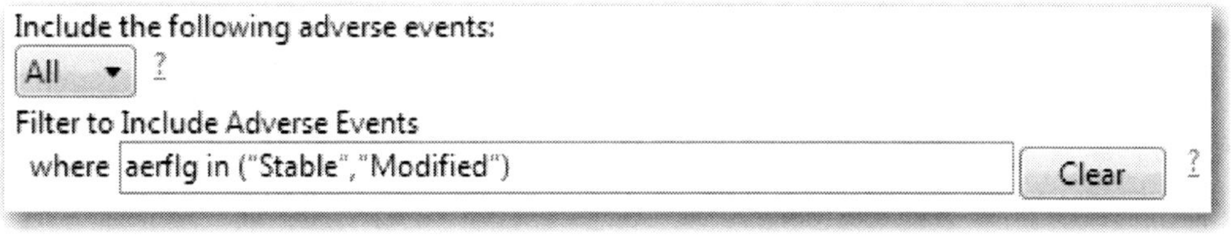

6.6 Final Thoughts

In this section, I summarized the snapshot comparison feature which can accelerate clinical reviews from one snapshot to the next. This feature is also useful for biostatisticians as they are preparing to lock the study database. Too frequently in every database lock, outstanding queries occur that must be tracked in the DBMS and confirmed within the study database. The snapshot comparison feature makes it straightforward for the study statistician to substantiate that the necessary changes have been implemented correctly, while at the same time provides reassurance that no other unexpected changes have occurred to the database.

Currently, record-level snapshot comparison features are surfaced in most Findings, AE, Event, and Intervention reports as well as **DM Distribution**, either by providing record-level review flags in the report **Data Filters**, or by having a **Include the following...** option on the **Filter** tab of report dialogs to easily subset analyses to New records. These record-level features are not available for **Risk-Based Monitoring** or **Fraud Detection** menus. However, as was done in "6.3.2 Patient-Level" on page 214, users can create subject-level filters "6.3.2 Patient-Level" on page 214to subset **Fraud Detection** analyses to those patients with New or Modified data.

As I mentioned in the introduction, the current implementation of the snapshot comparison feature does not consider data from the Trial Design domains or Relationship datasets (i.e. SUPP domains or RELREC). For Trial Design domains, this is probably less of a concern since these data are supplied wholly by the trial sponsor; comparisons for data in SUPP domains and RELREC are certainly more important. From my experience, SUPPxx domains often only include leading questions or domain-specific investigator comments, so the careful tracking of these data from snapshot to snapshot may be less of a concern. However, should users want to utilize the snapshot comparisons feature for their SUPPxx data sets, they can include the appropriate SUPPxx variables within their SDTM domains for analysis within JMP Clinical. Snapshot comparisons for SUPPxx domains will be implemented in a future version.

References

1 Haley EC, Kassell NF and Torner JC. (1993). A randomized controlled trial of high-dose intravenous nicardipine in aneurysmal subarachnoid hemorrhage. *Journal of Neurosurgery* 78: 537-547.

2 CDISC Submission Data Standards Team. (2013). *Study data Tabulation Model Implementation Guide: Human Clinical Trials, Version 3.2*. Round Rock, TX: Clinical Data Interchange Standards Consortium.

7

Final Thoughts

A Work in Progress 239
Stay in Touch 240

A Work in Progress

In this book, I summarized several methodologies to enable the centralized monitoring of clinical trials, using an example cardiovascular study to illustrate concepts. Methods include the risk-based monitoring (RBM) approaches as outlined by TransCelerate BioPharma Inc., numerous statistical and graphical techniques to identify quality issues or fraud, and comparisons between data snapshots to easily identify new or modified data as the trial progresses. Further, I describe the implementation of these tools in terms of the CDISC domains and variables that are required in order to perform the various analyses. This last point is particularly important. Though the terms "CDISC" or "data standards" do not appear anywhere in the title of this book, they are every bit as important as the statistics or visualizations presented for the efficient review of clinical trials. I look forward to the day when these standards are required for regulatory submissions. Not only will these data standards streamline the regulatory review process and the interactions between various entities involved in pharmaceutical development, they will enable more straightforward meta-analyses of the safety and effectiveness of treatments across clinical trial sponsors and time.

I made notes describing the future development of potential new features within JMP Clinical at the conclusion of each chapter. Some developments are straightforward for completeness of presentation, such as including additional reports for the fraud detection techniques cited, though not yet implemented. Other enhancements will require user feedback on how best to build rich and informative reviews that anticipate the appropriate follow-up questions for completeness. Some features simply have to wait for all of the necessary pieces to be in place in order to make them available to users. Perhaps the best example of this last point involves the prediction of quality or safety problems so that sponsors can intervene before they worsen to unacceptable levels. JMP Clinical and JMP Genomics already have sophisticated predictive modeling capabilities specifically tailored to address the needs of life science applications; work remains to be done to integrate these models for straightforward implementation within the RBM reports.

In the meantime, there is sufficient functionality available to greatly improve upon how clinical reviews are currently being conducted. To take it a step further, JMP Clinical is software that can be used by and offers something for every member of the clinical trial team, representing a new paradigm for efficient drug development. Embracing modern drug development will take time; employees will need training, new standard operating procedures will need to be written and adopted, methods will need development and refinement, and new systems will need to be put in place. As the Chinese philosopher Lao-Tzu once said "a journey of a thousand miles begins with a single step".

Stay in Touch

While I rely heavily on my past experience within the pharmaceutical industry to guide how new methods are implemented within JMP Clinical, I need your feedback so that the answers you need from your data are within easy reach. In particular, please feel free to share your professional experiences. While there is some literature on methodologies to identify misconduct and quality issues in clinical trials, I suspect the literature is incomplete due to potential fears of admitting such problems exist, or the perceived risk that documenting the experience may pose to your company. If you are unable to publish a particular case study, consider getting in touch. Even a conversation held in the most general of terms may lead to something useful. After all, every new method for detecting problematic data made available within JMP Clinical will greatly serve patients and the clinical trial community.

Feel free to connect on LinkedIn, follow me on Twitter (@rczink), or send emails directly to SAS Publishing at saspress@sas.com.

Recommended Reading

- *JMP Essentials, Second Edition*
- *Implementing CDISC Using SAS: An End-to-End Guide*
- *Jump into JMP Scripting*
- *A Guide to Statistics and Data Analysis Using JMP, Fifth Edition*

For a complete list of SAS books, go to support.sas.com/bookstore. If you have questions about which titles you need, please contact a SAS Book Sales Representative:

SAS Books
SAS Campus Drive
Cary, NC 27513-2414
Phone: 1-800-727-3228
Fax: 1-919-677-8166
E-mail: sasbook@sas.com
Web address: support.sas.com/bookstore

Recommended Reading

Index

Special Characters

JMP Essentials, Second Edition 12
JMP Start Statistics: A Guide to Statistics and Data Analysis Using JMP, Fifth Edition 12
Jump into JMP Scripting 12

A

actions
  default 48
  defining alternate 80
  defining for elevated risk 93
ADaM 8
Add Command 111
Add Notes drill down 59, 67, 217
Add RBM Variables command 112
Add Study from CDI 75
Add Study From Folders 14, 75
Add Study from Metadata Libraries 75
Add Study reports 207
Add Variable
  drill down 34, 38
Add Variable drill down 75, 93
Add-In menu 109, 112
Adverse Events (AE) 25, 78, 114
AE Distribution 232
AE Narratives 26, 70
Analysis notes 218
Analyze menu 95, 99
Arrow tool 191

B

bar charts 103
Benjamini, Y. 138
Between-Subject Distance Summary tab 190
Big Class data table 11
Big Class distribution output 11, 12
Birthdays and Initials report 181, 194, 195

C

case report forms (CRFs) 20
Category variable 85

CDI (Clinical Data Integration) product 10
CDISC (Clinical Data Interchange Standards Consortium) 8, 9
CDISC-formatted data sets 22
Center Flag variable 83, 84, 86
Center Value variable 82, 83, 84, 86, 90
Check Required Variables 12
Check Risk Threshold data set 84
Clinical Data Integration (CDI) product 10
Clinical Data Interchange Standards Consortium (CDISC) 8, 9
clinical sites
 hierarchical clustering of subjects within 168
Clinical Starter menu 9
clinical trials
 about 3
 examples of 13
Cluster Subjects Within Study Sites report 169
Clustering dialog 95, 99
Cochran-Mantel Haenszel (CMH)
 row mean score statistic 138
Color Clusters 96
Comment variable 87
Computed Deviations metric 34
Computed Eligibility Violations variable 38, 39, 42
Computed Eligibility Violations variable 53, 54, 87, 100
Constant Findings report 144, 197
contact information 240
CONTENTS procedure 203
Country notes 219
Country-Level Distributions tab 67, 68
country-level risk 66
Country-Level Risk Indicators data table 42
Country-Level Risk Indicators data table 67, 68, 92, 109
Create Subject Filter drill down 216
CRF Entry Response Time metric 34, 67
CRFs (case report forms) 20

D

Data Filter 130, 132, 135, 149, 150, 155, 158, 160, 182, 214, 217, 225, 232, 234
Data Filters report 237
data sets
 CDISC-formatted 22
 Check Risk Threshold 84
 Default Risk Threshold 42, 53, 57, 75, 79, 80, 84, 87, 90, 93, 95, 103, 121

Define Risk Threshold 21, 48, 53, 75, 80, 83
Define Study Risk 87
Manage Risk Threshold 21
Risk Threshold 24, 67, 76, 87, 88, 90, 93
Save Risk Threshold 83
Study Risk 34, 35, 36, 37, 39, 42, 52, 75, 87
Update Study Risk 21, 28, 29, 31, 39, 49, 53, 54, 57, 75, 76, 85, 87, 89, 93, 100, 108
data sources, risk indicators from other 31
data standards
  importance of 8
data tables
  Big Class 11
  Country-Level Risk Indicators 42, 67, 68, 92, 109
  Lock 225
  Monitor Summary 122
  Risk Indicator 53
  Site-Level Risk Indicators 42, 92, 95, 100, 103, 104, 106, 109, 110, 117, 121
default actions 48
Default Risk Threshold data set 42
Default Risk Threshold data set 53, 57, 75, 79, 80, 84, 87, 90, 93, 95, 103, 121
default risk thresholds 44
Define Risk Threshold data set 21, 48, 53, 75, 80, 83

Define Study Risk data set 87
Demographic Distribution 225
digit preference 154
Direction for Risk Signals variable 82, 83, 84, 86, 90
discontinuation variables 108
DISTANCE procedure 169
Distribution Details tab 225
Distribution dialog 11
distributions 60
domain keys 200
Domain Viewer 220, 221
drill downs
  Add Notes 59, 67, 217
  Add Variable 34, 38, 75, 93
  Create Subject Filter 216
  Export Blank Tables 34
  Export Tables 34
  Import Tables 34, 35
  Open Edit Checks 26, 72
  Report Actions 56, 86
  Risk Indicator 54, 56, 60, 64, 67, 95, 96, 100, 106, 107
  Select Rows Using Risk Indicators 56, 59, 68
  Select Sites Using Country Selection 68
  Show Rows in Heat Map 172, 191
  Show Subjects 26, 70, 174, 181, 182, 190
  Subset Clustering 174, 191
  View Domain 221
  View Notes 59, 67, 217, 224, 229

Visit Bar Chart 139
Duplicate Records report 148, 153
duplicate sets
 of measurements 148
Duplicates and Keys report 205, 207, 208, 209
DV (Protocol Deviations) 25

E

electronic case report forms (eCRFs) 2
examples
 of clinical trials 13
Expected Randomized 83
Export Blank Tables
 drill down 34
Export Tables
 drill down 34
Exposure (EX) domain 25, 76

F

False Discovery Rate (FDR) 138
Filters tab 71, 139, 160, 165, 171, 216, 232, 236
Findings Time Trends report 157
Fit Model platform 124
For Each Row function 120
fraud detection
 about 5, 127
 digit preference 154
 measurements collected at clinical site 144
 multivariate analyses 164
 plots of summary statistics by site 161
 study visits 129
 time trends 157
Function
 menu 39

G

General tab 71, 236
GEOCODE procedure 49
Geocode Sites 49, 51
Graph Builder 103
Graph menu 95
graphing
 bar charts 103
 maps 106

H

Help menu 12
hierarchical clustering
 of pre-dosing covariates across clinical sites 185
 of sites 95
 of subjects within clinical sites 168
Hochberg, Y. 138

I

Import Tables
  drill down 34, 35
Inclusion/Exclusion Criterion
  Not Met (IE) 25
individual risk indicators 22, 80
Integrated Quality Risk
  Management Plan (IQRMP)
  20, 48
Interactive Voice Response
  System (IVRS)
  logs 31
Interactive Voice Response
  System (IVRS) logs 31
International Conference on
  Harmonisation (ICH) 3, 19

J

Jacknife Distances 102
JMP add-ins
  creating 108
JMP Clinical 9
JMP scripts
  creating 108

K

KMeans option 99

L

Label variable 85
Laboratory Test Results tab
  145
last-patient-last-visit (LPLV)
  status 7
Local Data Filter 56, 63, 64,
  190
Lock data table 225
locking
  the trial database 7

M

Magnifier tool 191
Mahalanobis Distance 100, 166
Manage Risk Threshold Data
  Set 21
Map Geocoding Help 21, 49
maps 64, 106
measurements
  collected at clinical site 144
  duplicate sets of 148
Missing Data tab 166
Missing Informed Consent 83
Missing Pages
  metric 34
Monitor 37
monitor level, analyses at 120
Monitor Summary data table
  122
mosaic plots 132
Multidimensional Scaling 99

multiplicity adjustments 138
multivariate analyses 164
Multivariate and Correlations platform 100
Multivariate Inliers and Outliers report 165
multivariate inliers/outliers 164

N

Name function 122
National Heart, Lung, and Blood Institute 128
National Surgical Adjuvant Breast and Bowel Project 127
Natural Key 202
NCluster Handle 96
New Script icon 109
Nicardipine clinical trial 13
notes, adding/viewing 217

O

Observed Minus Expected Randomized 83
Open Differences Report button 210, 214
Open Duplicates and Keys Report button 201
Open Edit Checks drill down 26, 72
Output Folder 99

Overall Risk Indicator 56, 59, 60, 64, 67, 68, 89, 100, 104, 106
Overall Risk Indicator Adverse Event 90, 96, 99
Overall Risk Indicator Manually Entered 43, 90, 102
overall risk indicators 42, 80, 89
Overdue Queries metric 34

P

Partial Correlation Diagram 99
patient fraud detection
  about 179
  hierarchical clustering of pre-dosing covariates across clinical 185
  initials and birthdate matching 180
  quality and fraud 192
Patient notes 218
patient profiles 26, 70, 228
patient-level review flags 214
Percent Discontinued due to AE variable 110
Percent Discontinued due to Death variable 110
Percent Discontinued of Randomized Subjects 82
Percent Lost to Followup variable 110
pharmaceutical development 1

Points icon 162
Protocol Deviations (DV) 25
pseudo-code
  important terms in 76

Q

quality
  fraud and 192
Query Response Time metric 34, 67

R

Record notes 219
record-level review flags 212
Red Magnitude variable 81, 82, 84, 86
Red Percent of Center variable 80, 81, 82, 84, 86
Red Recommended Action variable 86
Report Actions drill down 56, 86
reports
  Add Study 207
  Birthdays and Initials 181, 194, 195
  Cluster Subjects Within Study Sites 169
  Constant Findings 144, 197
  Data Filters 237
  Duplicate Records 148, 153
  Duplicates and Keys 205, 207, 208, 209
  Findings Time Trends 157
  Multivariate Inliers and Outliers 165
  Weekdays and Holidays 129, 132, 133, 135
Revert Clustering 191
review flags
  Demographic Distribution 225
  Domain Viewer 220
  patient profiles 228
  patient-level 214
  record-level 212
  using 220
risk
  country-level 66
  defining actions for elevated 93
  site-level 53
  subject-level 70
  thresholds 108
Risk Indicator data table 53
Risk Indicator drill down 54, 56, 60, 64, 67, 95, 96, 100, 106, 107
risk indicators
  definitions of 76
  individual 22
  overall 42
  site-level risk 53
risk indicators country-level risk 66
risk indicators subject-level risk 70

risk threshold 44
Risk Threshold data set 24, 67, 76, 87, 88, 90, 93
risk thresholds 80, 108
risk-based monitoring (RBM)
    about 3, 19, 79
    Adverse Event variables and figures 114
    analyses at monitor level 120
    creating JMP scripts and add-ins 108
    defining actions for elevated risk 93
    defining alternate risk thresholds and actions 80
    dialog 23
    geocoding sites 49
    graphing 103
    menu 21
    report 24
    reports 75
    risk indicators 22, 53, 76
    statistical analyses 95
    weights for overall risk indicators 89
risk-based monitoring (RBM)report 106
Run icon 110
Run Script button 11

S

SAEs (serious adverse events) 20

Save Risk Threshold data set 83
Scatterplot Matrix 100
SDTM (Study Data Tabulation Model) 8
SDV (source data verification) 3, 19
Select Rows Using Risk Indicators drill down 56, 59, 68
Select Sites Using Country Selection drill down 68
serious adverse events (SAEs) 20
Share Report button 182
Show Control Panel 104
Show Domain Keys 224
Show Dropped Records 224
Show Duplicates 224
Show Rows in Heat Map drill down 172, 191
Show Subjects drill down 26, 70, 174, 181, 182, 190
Signal variable 54
Site 39 Distance Matrix tab 172
Site Active Date 34, 37
Site notes 220
Site-Level Distributions tab 56, 60
site-level risk
    distributions 60
    maps 64
    Risk Indicator Data Table 53
Site-Level Risk Indicator Maps tab 52, 64

Site-Level Risk Indicator tab 54, 59
Site-Level Risk Indicators
 data table 42
Site-Level Risk Indicators data table 92, 95, 100, 103, 104, 106, 109, 110, 117, 121
sites hierarchical clustering of 95
sites identifying extreme sites using Mahalanobis Distance 100
sites plots of summary statistics by 161
Smoother 106
snapshot comparisons
 about 7, 199
 adding notes 217
 AE Distribution 232
 domain keys 200
 record-level review flags 212
 review flags 212, 220
 viewing notes 217
SOPs (Standard Operating Procedures) 2
SORTEDBY metadata 205
source data verification (SDV) 3, 19
Specify Target Enrollment 54
Standard Operating Procedures (SOPs) 2
statistical analyses
 hierarchical clustering of sites 95

Study Data Tabulation Model (SDTM) 8
Study Risk
 data set 34, 35, 36, 37, 39, 42
Study Risk data set 52, 75, 87
Study Site Identifier variable 104
study visits
 study scheduling 136
 weekdays and holidays 129
subject-level risk 66
Subset Clustering drill down 174, 191
summary statistics 104, 161
Surrogate Key 202

T

T Square 102
Tables tab 182, 183, 228
tabs
 Between-Subject Distance Summary 190
 Country-Level Distributions 67, 68
 Distribution Details 225
 Filters 71, 139, 160, 165, 171, 216, 232, 236
 General 71, 236
 Laboratory Test Results 145
 Missing Data 166
 Site 39 Distance Matrix 172
 Site-Level Distributions 56, 60

Site-Level Risk Indicator 54, 59
Site-Level Risk Indicator Maps 52, 64
Tables 182, 183, 228
Total Signs 148
VS Treatment Time Trends 158
Tabulate platform 121
tests
 with no variability 144
thresholds
 for individual risk indicators 80
 for overall risk indicators 80
 for user-added risk 87
 risk 108
time trends 157
Total AEs on Study 27
Total Queries metric 34
Total Signs tab 148
Two Way Clustering 99

U

Update Study Risk
 data set 31, 39, 49
Update Study Risk data set 21, 28, 29, 53, 54, 57, 75, 76, 85, 87, 89, 93, 100, 108
user-added risk
 thresholds for 87

V

V List Box() 119
variability
 tests with no 144
Variable variable 85
variables
 Category 85
 Center Flag 83, 84, 86
 Center Value 82, 83, 84, 86, 90
 Comment 87
 Computed Eligibility Violations 38, 39, 42, 53, 54, 87, 100
 Direction for Risk Signals 82, 83, 84, 86, 90
 discontinuation 108
 Label 85
 Percent Discontinued due to AE 110
 Percent Discontinued due to Death 110
 Percent Lost to Followup 110
 Red Magnitude 81, 82, 84, 86
 Red Percent of Center 80, 81, 82, 84, 86
 Red Recommended Action 86
 Signal 54
 Study Site Identifier 104
 Variable 85
 Weight for Overall Risk Indicator 80, 83, 84, 86, 89
 Yellow Magnitude 81, 82, 84, 85

Yellow Percent of Center 80, 81, 84, 85, 86
Yellow Recommended Action 86
View Domain drill down 221
View Notes drill down 59, 67, 217, 224, 229
Visit Bar Chart drill down 139
VS Treatment Time Trends tab 158

W

Ward method 95
Weekdays and Holidays report 129, 132, 133, 135
Weight for Overall Indicator 42, 43, 44, 89, 90
Weight for Overall Risk Indicator variable 80, 83, 84, 86, 89
WHERE statement 24, 25, 58, 77

Y

Yellow Magnitude variable 81, 82, 84, 85
Yellow Percent of Center variable 80, 81, 84, 85, 86
Yellow Recommended Action variable 86

Z

Zink, Richard C. 240

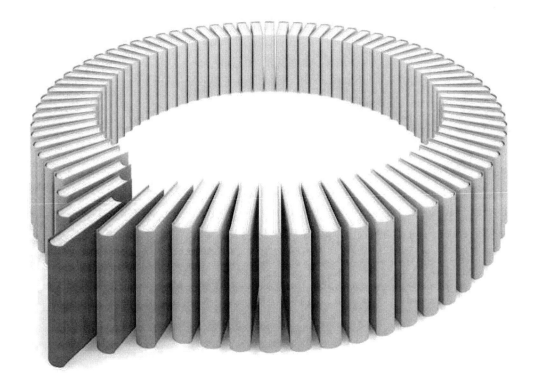

Gain Greater Insight into Your SAS® Software with SAS Books.

Discover all that you need on your journey to knowledge and empowerment.

CPSIA information can be obtained at www.ICGtesting.com
Printed in the USA
LVOW10s2338300814

401587LV00003B/18/P